SpringerBriefs in Electrical and Computer
Engineering

More information about this series at http://www.springer.com/series/10059

Ruonan Zhang • Lin Cai • Jianping Pan

Resource Management for Multimedia Services in High Data Rate Wireless Networks

 Springer

Ruonan Zhang
School of Electronics and Information
Northwestern Polytechnical University
Xi'an, Shaanxi, China

Lin Cai
Department of Electrical and Computer
  Engineering
University of Victoria
Victoria, British Columbia, Canada

Jianping Pan
Department of Computer Science
University of Victoria
Victoria, British Columbia, Canada

ISSN 2191-8112   ISSN 2191-8120   (electronic)
SpringerBriefs in Electrical and Computer Engineering
ISBN 978-1-4939-6717-9   ISBN 978-1-4939-6719-3   (eBook)
DOI 10.1007/978-1-4939-6719-3

Library of Congress Control Number: 2016953468

Printed on acid-free paper

This Springer imprint is published by Springer Nature
The registered company is Springer International Publishing AG
The registered company address is: Gewerbestrasse 11, 6330 Cham, Switzerland

Contents

Chapter 1
Overview

Wireless local area networks (WLANs) have swept across the globe in recent years as an inexpensive and convenient means for ubiquitous Internet access, thanks to the popularity of devices such as smart phones and tablet computers. By wireless communications and the Internet backbone technologies, WLANs provide services to both mobile and fixed users and have enabled a variety of applications. Given the lower cost and relatively higher throughput compared to cellular networks, WLANs are increasingly used to provide multimedia services such as video streaming. However, multimedia services have posed new requirements and significant technical challenges on the treatment of traffic within WLANs, to provide users seamless coverage and quality-of-service (QoS). The resource management of WLAN has become a very crucial part for future generation of wireless and mobile networks. Nevertheless, QoS is only partially supported in current WLANs, which has become an obstacle to supporting multimedia applications nowadays. In this chapter, we present an overview of the essential concepts of QoS, fundamental principles and challenges of QoS provisioning, and the resource management mechanisms for WLANs.

1.1 Background

The wireless and mobile Internet is quickly emerging as a reality, thanks to the exponential rise in the usage of mobile devices and the fast evolution of the wireless local area network (WLAN) technologies. WLANs provide the same functionality as wired LANs, but with much better flexibility and mobility without physical location constraints and the costs of installing wires. WLANs have been not only widely installed in hot-spot locations, from markets, airports, and campus to retail, hospital, and corporate buildings, but also in residential buildings to support

© The Author(s) 2017
R. Zhang et al., *Resource Management for Multimedia Services in High Data Rate Wireless Networks*, SpringerBriefs in Electrical and Computer Engineering, DOI 10.1007/978-1-4939-6719-3_1

Internet access. WLANs can also be utilized in places that are difficult to wire, for example, trading floors, manufacturing facilities, and warehouses. Hence, WLANs have become ubiquitous and an integral part of the Internet.

Given the license-free usage of spectrum and high data rates, the main usage of WLANs is no longer limited to data services such as Web browsing, file transfer, and Emails. Convergence of WLANs and the Internet has promoted a variety of applications for mobile users including real-time multimedia services such as the delivery of video streaming or video on demand (VoD), voice telephony or voice over IP (VoIP), teleconferencing, and instantaneous monitoring. There are mainly two driving forces. First, network service providers are racing to roll out new value-added services over the Internet, such as Internet protocol television (IPTV). Second, consumers demand for multimedia streaming to their handheld devices. For example, when using WiFi and cellular-enabled dual-mode phones, people more likely switch to WiFi connectivity in the range of an access point (AP) for much lower costs and higher data rates. As a result, recent years have seen explosive growth of VoIP over WLAN, where compressed voice data are encapsulated into IP packets and transferred to the subscribers through the Internet instead of the traditional public switched telephone network (PSTN).

Multimedia applications often have stringent quality-of-service (QoS) requirements on throughput, delay, delay jitter, and delivery ratio. For example, with the state-of-the-art video coding technologies, the average data rates of high-quality, high-definition (HD) video streams are reduced significantly, but the more aggressive source coding and compression lead to a much higher burstiness (the peak-to-average ratio) in video streaming traffic [128]. These features lead to the great difficulty in handling multimedia traffic.

A mechanism that provides a certain level of QoS in wireless networks is usually termed as a *QoS enabling mechanism* or *QoS support mechanism*. QoS provisioning refers to the ability of a network to ensure data delivery over the network and satisfy certain performance requirements for different classes of traffic. For wired links, an engineering solution, bandwidth over-provisioning, is widely used to address the QoS issues. For example, it can be achieved by using high-speed, reliable fibers. However, the wireless spectrum is at premium and the demand for wireless services is ever-increasing, and thus bandwidth over-provisioning is not a sustainable solution in wireless networks. Furthermore, the physical channel impairments such as fading and interference, super-linearly decaying of signal strength, and hidden-terminal issues make QoS support much harder. Third, to support multiple users simultaneously, WLAN resources are usually accessed in a contention-based and distributed manner, given the broadcast nature of wireless channels. These characteristics hence render effective and efficient resource management in WLANs crucial and challenging. A properly designed resource management scheme is the key to efficiently allocate radio resources and ensure QoS.

1.2 Wireless Resource Management Basics

Resource management has attracted significant interest and research efforts in both the academical and industrial communities. The key issue is how to utilize the limited wireless resources efficiently and fairly. Two aspects arise when dealing with the resource management with QoS constraints. One is the call admission control (CAC), which determines whether it is possible to fulfill the demands of a set of clients (i.e., calls or flows). The other is to find an optimal way to share/schedule the resources among multiple competing stations, in order to meet the demands from all clients. In a distributed system, the latter is realized by the media access control (MAC) protocols which coordinate users to utilize a common communication channel in a network.

In WLANs, CAC normally operates at APs or gateway devices, and works on the call or session level. It makes the decision on whether or not an incoming call (or a session) can be accommodated such that the QoS requirements of the admitted calls can be statistically guaranteed [5, 26, 144]. MAC operates at each wireless station and it controls the channel access on the link-layer frame level. Aiming to fairly share the common wireless resources and avoid network congestion, each station decides when to transmit in a distributed manner or is permitted to transmit by a central controller/scheduler.

In a practical wireless network, resource management should be feasible and with low overhead and computational complexity. Overheads are mainly introduced by the additional fields defined in MAC protocols, and the exchange of signaling messages. Effective QoS mechanisms must make a good tradeoff between the introduced overhead and the efficiency in utilizing network resources [101]. In this monograph we focus on the modeling and performance analysis of the WLAN MAC protocols, which serves as a foundation for their performance improvement and optimization.

Resource management by MAC can take different approaches. In the following, we briefly introduce and compare the three major streams for wireless networks.

(1) Centralized Resource Allocation

To ensure QoS for multimedia applications, wireless resources can be fully or statistically reserved before a new call is admitted to a network. In the cellular systems, wireless resources in the time/frequency/code domains can be channelized. Each admitted call is assigned with a dedicated channel using one or multiple time slots, frequency bands (subcarriers), and codes. Correspondingly, the fundamental resource allocation for multiple users includes the time-division multiple access (TDMA), frequency-division multiple access (FDMA), and code-division multiple access (CDMA). Reference [2] discussed the centralized resource allocation strategies used in the third-generation (3G) system.

(2) Distributed Resource Reservation

Centralized resource allocation is often associated with the single point of failure vulnerability, high overhead, and high computational complexity. For flows with multimedia traffic, centralized resource allocation becomes even more expensive and difficult. Thus, distributed reservation has been proposed and adopted in some standards for LANs, where each wireless station sends channel reservation requests in its own beacons, and the successfully reserved channel will not be accessed by other competing stations. However, for the distributed reservation approach, there is no guarantee that the reservation is always successful when a network is congested.

(3) Contention with Service Differentiation

No matter centralized or distributed resource allocation or reservation, how many resources should be allocated/reserved for a flow with variable data rate is a challenging problem. Over-reservation may result in low resource utilization and under-reservation may result in unsatisfactory QoS performance. For bursty traffic, accessing the channel using a fair contention mechanism is preferable. IEEE 802.11 standards define a set of MAC and physical layer (PHY) specifications for implementing WLANs [124], and the distributed coordination function (DCF) of 802.11 has been widely adopted. To support heterogeneous applications with different QoS requirements, the enhanced DCF (EDCF) has been proposed and adopted in the IEEE 802.11e standard [54], where the system parameters such as the interframe space (IFS), contention window (CW) size, and backoff interval (BI) can be adjusted for service differentiation. However, without resource reservation, the prioritized contention using EDCF cannot ensure hard QoS provisioning for multimedia services.

1.3 Multimedia Traffic Characteristics and Performance Metrics

Different multimedia applications have different traffic characteristics, which directly affect the network queueing behavior and system performance. Since 2015, video traffic on mobile devices accounts for more than 50 % of the total mobile data [28], so we focus our attention on video applications here. The H.264/AVC (MPEG-4 Part 10) video codec standard has been widely used to transmit and store HD video content thanks to its high compression efficiency. Video streams have a constant refresh rate, typically 30 frames per second, and frames are encoded using three encoding techniques resulting in three types of video frames, I, P, and B frames. I frames are encoded independently and have larger frame sizes, while P and B frames are encoded by taking reference to other frames and thus have much

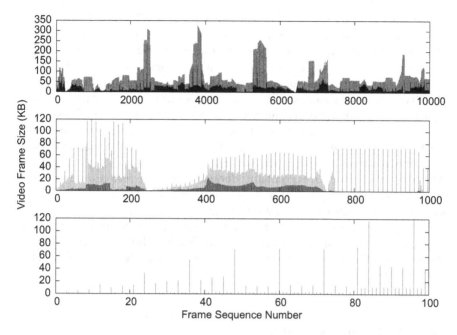

Fig. 1.1 Video frame size vs. frame sequence number [155]

smaller frame sizes, as shown in Fig. 1.1 for the sample video stream.[1] A number of frames are grouped as one group of picture (GoP), e.g., in H.264, each GoP has 12 frames with the structure of "IBBPBBPBBPBB".

Since a limited size of maximum transmission unit (MTU) is used in the link layer, which is smaller than most video frames, video frames are segmented and encapsulated into multiple link-layer frames. Because the inter-frame interval is fixed ($1/30$ s), the dramatic variation of video frame sizes within a GoP and between GoPs results in the high burstiness of video traffic. For the sample video stream, the peak-to-average ratio of the instantaneous data rate can be as high as 16.18.

On the other hand, multimedia applications and mobile devices have a variety of service requirements that should be considered when designing effective resource management solutions. The main QoS metrics for multimedia applications include delay, jitter, throughput, and loss [37, 71]. *End-to-end frame delay*, the time from the arrival of a frame at its source station to the moment it is received by its destination station, is a key QoS index in WLANs. The waiting time in the source station buffer (queueing delay) and the backoff time for obtaining transmission opportunities till being successfully transmitted (channel access delay) for a frame

---

[1]The sample HD video stream is "From Mars to China" with a resolution of 1920 × 1080 and quantization parameters of 28, 28, and 30 for I, P, and B frames, respectively. The video is available at http://trace.eas.asu.edu/h264/mars/.

are the main components of the end-to-end delay in a WLAN. The other delay components, such as processing time, propagation delay, and reception time are relatively small, so they can be negligible. The end-to-end delay directly affects the user-perceived service quality. The variation of delay is called *delay jitter*. *Throughput* is related to the data transfer rate from the source to the destination which is defined as the amount of data successfully received in a unit time. Packets (a packet may be segmented into multiple frames) may be lost during transmissions due to various reasons, such as transmission error, buffer overflow, collision, etc., which is evaluated by *packet loss rate*. The International Telecommunication Union-Telecommunication Standardization Sector (ITU-T) G.114 [71] has specified the QoS performance indexes. For instance, for real-time voice services, the tolerable packet loss rate for raw voice data is 13 %. The delay below 150 ms is preferable for one-way transmission, but it should not be larger than 400 ms.

For portable devices powered by batteries, such as smart phones and tablet computers, power consumption and computation load are also important and should be considered. Moreover, being adaptive to wireless channel dynamics is critical for wireless networks [6].

1.4   Challenges and Issues

QoS support in WLANs has made significant progress by extensive research efforts, but providing satisfactory services for multimedia applications in wireless networks still poses a number of challenges as listed below.

First, it is difficult to provide hard QoS guarantee when distributed channel access control is employed. To fairly share wireless resources and avoid network congestion collapse, DCF uses a backoff counter randomly set according to the backoff window size. A wireless station reduces its transmission rate by enlarging the backoff window size exponentially when a frame loss is observed in the first several backoff stages (the contention window size may be fixed when the number of retransmissions has reached a threshold). However, using exponential backoff and random backoff counters introduces variations in delay and throughput. When more than one node access the channel, frame collisions introduce more delay and loss, resulting in more difficulties in QoS guarantee. Even with prioritized contention using EDCF, only soft QoS provisioning can be realized. How to support real-time multimedia streaming with strict QoS requirements in WLANs remains an open issue.

Second, multimedia traffic is multiplexed with data traffic in WLANs, different from the cellular networks where dedicated resources are allocated to multimedia flows. Consequently, the QoS of real-time applications may be severely degraded when the best-effort data traffic load increases. It is a challenging task to guarantee QoS provisioning for multimedia flows and also try to ensure high throughput for best-effort flows at the same time.

Third, as more wireless devices, such as WiFi, Bluetooth, and ZigBee devices, access the same unlicensed wireless spectrum, packet losses due to interference and collisions become a challenging issue to deal with. Furthermore, the mobility of portable devices introduces channel fading and random variations of the received signal-to-noise ratio (SNR). Since wireless transmission links are error-prone, highly unreliable, and time-varying, retransmissions of lost packets are usually required. How to optimize the channel sharing among retransmitted and new packets should be studied.

Fourth, using adaptive modulation and coding, the physical layer of wireless networks can support multiple data rates (e.g., 1, 2, 5.5, and 11 Mb/s in IEEE 802.11b) according to the channel condition. The rate adaptation introduces more delay and throughput variations that should be considered in resource management.

In summary, effective and efficient QoS support mechanisms and appropriate MAC parameter setting for both multimedia traffic with stringent QoS requirements and best-effort data are still an open and challenging issue, which introduces rich research topics.

1.5 Monograph Outline

The monograph consists of five chapters. It begins with the introduction of the QoS metrics, and the basics and challenges of the resource management in WLANs, as presented in this chapter.

The wireless network specifications defined by the standardization bodies such as IEEE are summarized in Chap. 2. The evolution, features, and application scenarios of the standards for WLANs, wireless personal area networks (WPANs), and wireless body area networks (WBANs), which are driven by the ever-increasing demand for services and the state-of-the-art radio and resource management technologies, are introduced. The survey provides readers with the useful background information of the primary MAC protocols and QoS support mechanisms in today's wireless networks.

In Chap. 3, the protocols, performance evaluation, and analytical models of the contention-based MAC are discussed. The fundamental theorems and the widely adopted analytical models (including the two-dimensional Markov chain, mean value analysis, and backoff counter distribution analysis) for both the ordinary and the prioritized contention-based MAC are presented. A taxonomy of the analysis approaches is provided, and the principles, concepts, and features of each approach are discussed.

The reservation-based MAC is investigated in Chap. 4. The reservation can be realized in two ways, i.e., centralized and distributed resource allocation. Using the WiMedia MAC protocol for high-rate WPANs [49] as an example, the distributed resource reservation is studied in details. The scheme, analytical model, and performance evaluation are presented. Furthermore, the improvement of the current protocol is discussed.

By combining reservation and contention-based media access together, the hybrid approach is promising for achieving a good tradeoff between QoS guarantee and resource utilization. Chapter 5 presents the design, analytical model, and performance evaluation of the hybrid-MAC. Both the hybrid approaches with hard and soft-reservation are investigated. Moreover, a case study of IPTV delivery in a home network using the hybrid-MAC is discussed.

Overall, this monograph introduces the principles in the QoS provisioning in wireless networks and the related link-layer resource management techniques, with the emphasis on protocol modeling and performance evaluation. The presented analytical studies and simulation results of the MAC protocols can help readers to understand and evaluate the network performance. Meanwhile, the physical-layer transmission techniques and the MAC protocols of the main standards of WLANs, WPANs, and WBANs are briefly reviewed, and their histories and key characteristics are summarized. A selected bibliography is provided at the end of the monograph for more in-depth reading and deeper understanding. Therefore, this monograph can serve as a knowledge source and reference for engineers, researchers, students, and users of wireless networks.

Chapter 2
MAC Protocols for High Data-Rate Wireless Networks

The ever-growing demand for multimedia services anywhere, anytime has fostered the development of high data-rate wireless networks, and various standards have emerged. In this chapter, we briefly introduce three categories of wireless networks, the wireless local area networks (WLANs), wireless personal area networks (WPANs), and wireless body area networks (WBANs). Their coverage scales down from over a building to a house/home and finally to a body. Their resource management and MAC protocols have distinct features due to different data rates, capacity, radio ranges, and application scenarios. The enabling physical-layer radio frequency (RF) transmission technologies, MAC schemes, and standardizations of the three kinds of networks are summarized, with the objective to provide readers with a preliminary understanding of the design philosophies and implementation requirements.

2.1 Wireless Local Area Networks

(1) Overview of IEEE 802.11 WLANs

IEEE 802.11 is a set of PHY and MAC specifications to implement WLANs among terminals, with or without APs that are connected to wide area networks. The radio communications in the 2.4, 5, and 60 GHz frequency bands are utilized in the physical layer and a set of MAC protocols are defined for channel access and resource management. The IEEE Local and Metropolitan Area Networks (LAN/MAN) Standards Committee (IEEE 802) is responsible for creating and maintaining the PHY and MAC specifications.

The US Federal Communications Commission (FCC) opened the 2.4–2.5 GHz spectrum for individual non-licensed usage in the late 1980s. IEEE, the world's largest technical professional organization, recognized the need for a standard that

© The Author(s) 2017
R. Zhang et al., *Resource Management for Multimedia Services in High Data Rate Wireless Networks*, SpringerBriefs in Electrical and Computer Engineering,
DOI 10.1007/978-1-4939-6719-3_2

fulfilled the demand for wireless communications and networking infrastructures. Work began on creating such a standard in September 1990, and the first approved and adopted version of IEEE 802.11 was released in June 1997. When the work to develop IEEE 802.11 started, the goal was to develop inter-operable wireless products reaching a data rate of over 1 Mbps. Then the working group has made a series of 802.11 enhancements, such as the IEEE 802.11e for QoS support, 802.11ac for data rate as high as 1 Gbps, and IEEE P802.11ax for dense deployments (e.g., those in stadiums and shopping malls).

Due to the low-cost chipsets and the convenient setup, IEEE 802.11 has been widely deployed and it impacts our daily lives and industry. Nowadays, consumer electronics and portable devices such as laptops, tablets, TV sets, and smart phones are typically equipped with an IEEE 802.11 radio, often branded as "WiFi". The IEEE 802.11 technologies have been globally popular in providing wireless access to the Internet from offices, homes, airports, hotels, restaurants, trains, and aircraft.

IEEE celebrated the 25th anniversary of IEEE 802.11, which has become the standard for the world's premier WLAN products, on September 10th, 2015. The evolution of the IEEE 802.11 family has promoted technology improvement and enabled a wide range of applications. Today the IEEE 802.11 Wireless LAN Working Group continues to evolve the standard to support new applications such as smart grid and Internet of Things (IoT).

(2) Evolution of IEEE 802.11 Standards

The 802.11 family includes a series of half-duplex communication techniques based on the same basic protocol specified in the first IEEE 802.11 standard which was published in 1997 and clarified in 1999. The great success in the market and the perceived capacity limits of the basic 802.11 standard have driven a prolific technology improvements and extensions. The revisions are denoted by an alphabet set of amendments. The first widely accepted one was 802.11b that appeared in 2000 and was followed by the major standards of 802.11a, 802.11g, 802.11n, and 802.11ac. The other specifications in the family, using the alphabets such as from c to f, h, and j, are corrections to the previous standards or service extensions and amendments.

After the release of 802.11-1997, a lot of feedbacks have been received by the IEEE 802 working group that the compatibility among the products from different vendors was not supported well. For example, the default encryption scheme, called wired equivalent privacy (WEP), could not work among different products. Thus, a certification program to ensure the compatibility and inter-operability among the commercial products is needed. In order to solve the issue of compatibility among different vendors, the wireless ethernet compatibility alliance (WECA) was founded in 1999 and renamed as the WiFi alliance (WFA) in 2003. The WFA was formed as a trade association and the products certified by WFA hold the WiFi trademark. Nowadays almost every wireless product using the IEEE 802.11 air interface has the WiFi certification and is labeled by the brand.

The history of the evolution of IEEE 802.11 specifications and the amendments in progress are listed in Tables 2.1 and 2.2, respectively. The key standards and amendments are summarized as follows.

1. 802.11-1997 (Initial 802.11 Standard) [65]

   The first version, IEEE 802.11-1997 (802.11 legacy), specifies three solutions in the PHY layer: frequency hopping spread spectrum (FHSS), direct sequence spread spectrum (DSSS), and infrared PHY schemes. The first two schemes use the S-band radio frequency (RF) transmission, operating in 2.400–2.500 GHz (referred to as the 2.4 GHz band) which belongs to the industrial scientific medical (ISM) frequency band under the FCC Part 15 Rules and Regulations [3]. The last one uses the infrared band at 316–353 THz. It is defined in the standard that all the three PHYs provide a basic data rate of 1 Mbps and an optional 2 Mbps mode. However, the commercial products of the infrared PHY scheme in 802.11-1997 actually do not exist in the market.

   The spectrum is sub-divided into 14 channels and each channel spans 5 MHz. The center frequency of the first channel is 2.412 GHz. In addition, a spectral mask is specified in 802.11 and it regulates the power distribution allowed over the channels. According to the mask, the signals are required to be attenuated by a minimum of 20 dB compared with the peak amplitude of the power spectrum of a channel when the frequency is ±11 MHz away from the center of the band. Hence the bandwidth of the signal over the 802.11 air interface is effectively 22 MHz. As a result, the center frequencies of stations of two geographically overlapped 802.11 WLANs must be separated by at least four channels. In other words, for multiple WLANs to operate at the same location, the stations should use every fourth or fifth channel to avoid signal frequency overlapping.

2. 802.11a (OFDM Scheme in 5 GHz Band) [73]

   The PHY layer based on the orthogonal frequency-division multiplexing (OFDM) signaling method was originally described in the 1999 specification, but was later defined in the 2012 specification which used the 4.915–5.825 GHz (referred to as the 5 GHz band). By following the original standard (802.11 legacy), 802.11a adopts the same MAC protocol and frame format. However, 802.11a specifies transmission and reception at the data rates from 1.5 to 54 Mbps (the rate is higher if the error correction code is counted), which yields effective throughput up to 20–30 Mbps. The vendors began to ship 802.11a products in 2001 due to the development of the commercial radio devices working at the 5 GHz band. Nowadays the term "802.11a" is used by most WiFi products (interface cards and routers) to indicate inter-operability at the 5.8 GHz band with 54 Mbps data rate.

   The advantage of using the 5 GHz band is that there may be less interference because fewer other systems may operate in this band. However, this higher carrier frequency brings a disadvantage: signals are absorbed more severely by the solid objects such as walls in their propagation due to the smaller wavelength compared with 802.11b/g in 2.4 GHz. Consequently, 802.11a can only penetrate over shorter distance and provide a smaller effective coverage range.

Table 2.1 The IEEE 802.11 standard and its amendments

Number	Approval date	Title	Comment
802.11-1997	1997/6/26	IEEE Standard for Wireless LAN Medium Access Control (MAC) and Physical Layer (PHY) Specifications	Initial standard. 1 and 2 Mbps, 2.4 GHz RF and infrared (IR) standard
802.11-1999	1999/3/18	Part 11: Wireless LAN Medium Access Control (MAC) and Physical Layer (PHY) Specifications	Superseded by ISO/IEC 8802.11-1999
ISO/IEC 8802.11-1999	2003	IEEE Std 802.11-1999 (R2003)	International standard
802.11a	1999/9/16	High Speed Physical Layer in the 5 GHz band	54 Mbps OFDM PHY @ 5 GHz
802.11b	1999/9/16	Higher Speed PHY Extension in the 2.4 GHz Band	11 Mbps DSSS PHY @ 2.4 GHz
802.11b-cor1	2001/10/10	Corrigenda to IEEE 802.11b-1999	Clarify Amendment 2
802.11c	2001	Media Access Control (MAC) Bridges	Bridging in wireless bridges or access points, included in IEEE 802.1D standard
802.11d	2001/7/13	Operation in Additional Regulatory Domains	Allow devices to comply with regional requirements
802.11e	2005/9/22	MAC Enhancements	Support for QoS
802.11f	2003/8/4	Inter-Access Point Protocol Across Distribution Systems Supporting IEEE 802.11 Operation	Released as 802.11.1 and withdrawn by IEEE-SA Standards Board on 2006/2/3
802.11g	2003/6/12	Further Higher Data Rate Extension in the 2.4 GHz Band	54 Mbps OFDM PHY @ 2.4 GHz
802.11h	2003/10/10	Spectrum and Transmit Power Management Extensions in the 5 GHz Band in Europe	In Europe, 5 GHz devices must implement 802.11h
802.11i	2004/6/24	MAC Security Enhancements	MAC security enhancements, known as WPA and WPA2 from WiFi Alliance
802.11j	2004/9/23	4.9 GHz to 5 GHz Operation in Japan	Compliance with Japanese 5 GHz spectrum regulation
802.11k	2008/6/12	Radio Resource Measurement of Wireless LANs	Discover the best available access point
802.11ma	2007/3/8	802.11 Standard Maintenance Revision	Prepared for 802.11-2007 that supersedes 802.11-1999

(continued)

Table 2.1 (continued)

Number	Approval date	Title	Comment
802.11mb	2011/3/31	802.11 Accumulated Maintenance Changes	Second maintenance, prepared for 802.11-2012
802.11n	2009/10/29	Enhancements for Higher Throughput	Increase the maximum net data rate to 600 Mbps using MIMO, frame aggregation, etc. @ 2.4 and 5 GHz
802.11p	2010/7/15	Wireless Access for the Vehicular Environment	Car to car communication, closely related to IEEE 1609
802.11r	2008/7/15	Fast Basic Service Set (BSS) Transition	Permit continuous connectivity handoffs in a seamless manner
802.11s	2010/9/30	Mesh Networking	Transparent multi-hop operation
802.11t	2009/12/31	Recommended Practice for the Evaluation of 802.11 Wireless Performance	Develop 802.11.2, administratively withdrawn by IEEE-SA on 2006/2/3
802.11u	2011/2/25	Interworking with External Networks	Convergence of 802.11 and GSM
802.11v	2011/2/9	Wireless Network Management	Allow client devices to exchange information about network topology
802.11w	2009/9/30	Protected Management Frames	Increase the security of management frames
802.11y	2008/11/6	3650–3700 MHz Operation in USA	High powered equipment to operate using the 802.11a protocol on a co-primary basis
802.11z	2010/10/14	Extensions to Direct Link Setup (DLS)	AP independent DLS
802.11-2012	2012/3/29	New Release	Include Amendments k, n, p, r, s, u, v, w, y, and z
802.11aa	2012/5/29	Video Transport Streams	MAC enhancements for robust audio and video streaming
802.11ac	2013/12/11	Very High Throughput <6 GHz	Enhancements for >1 Gbps throughput below 6 GHz
802.11ad	2012	Very High Throughput 60 GHz	Enhancements for >1 Gbps throughput @ 60 GHz band
802.11ae	2012/4/6	Prioritization of Management Frames	Communicate management frame prioritization policy
802.11af	2013/11/11	TV White Spaces	Geolocation-based spectrum databases and channel sensing

Table 2.2 The IEEE 802.11 amendments in progress

Number	Approval date	Title	Comment
802.11mc	2016	Standard Maintenance	Roll-up of 802.11-2012 with the aa, ac, ad, ae & af amendments, prepared for 802.11-2016
802.11ah	2016	Sub 1 GHz band, Machine-to-Machine communications	Enhanced power saving mechanisms and efficient small data transmissions
802.11ai	2016/9	Fast Initial Link Setup	Reduce link setup time to below 100 ms
802.11aj	2016/6	China Millimeter Wave	Very high throughput WLAN using mmWave Ml MO @ 45 GHz
802.11ak	2016/5	General Links	Support IEEE 802.11 links for transit use in bridged networks
802.11aq	2016/7	Pre-association Discovery	Further discover the services running on a device or provided by a network
802.11ax	2019	High Efficiency WLANs	Dynamic channel bonding, multi-user uplink MIMO, and full-duplex wireless channel
802.11ay	TBD	Enhancements for Ultra High Throughput in and around the 60 GHz Band	Extension of 802.11ad to extend throughput, range, and use-cases by channel bonding, MIMO and higher modulation schemes
802.11az	TBD	Enhancements for Positioning	Enables determination of absolute and relative position with better accuracy

3. 802.11b (DSSS Scheme in 2.4 GHz Band) [74]

The 802.11b standard directly extended the modulation technique of the initial 802.11 in the PHY and employs the same media access method in the link layer. 802.11b adopts DSSS modulation and a channel has the bandwidth of 22 MHz, resulting in three "non-overlapping" channels (channel indexes of 1, 6, and 11). Compared with the 802.11-1997 which provides the mandatory data rate of only 1 Mbps, the maximum raw data rate in 802.11b is 11 Mbps. Due to the significant throughput increase and price reduction, 802.11b received rapid acceptance and became popular when it was released in 2000.

Meanwhile, other products operating in the 2.4 GHz ISM band, such as microwave ovens, cordless telephones, baby monitors, Bluetooth, ZigBee, and some amateur radio equipments, cause severe interference to 802.11b devices. In order to control the susceptibility, the DSSS scheme is adopted in the physical layer to mitigate the interference.

4. 802.11g (OFDM Scheme in 2.4 GHz Band) [75]

The 802.11g standard was ratified in June 2003 and it specifies the third modulation standard, i.e., the same OFDM-based scheme as 802.11a, but works in the 2.4 GHz band (same as 802.11b) with the channel bandwidth of 20 MHz. In the physical layer, besides the forward error correction codes, a maximum bit rate of 54 Mbps can be provided. Working in the same frequency band, the hardware of 802.11g is fully backward compatible with that of 802.11b. However, due to this legacy issue, the throughput of 802.11g is reduced by 21 % compared to 802.11a. In addition, when an 802.11g network and an 802.11b network co-exist, the data rate of the former will be reduced due to the activities of the latter. Similarly to 802.11b, 802.11g devices experience interference from other products that also operate in the 2.4 GHz ISM band, and the OFDM signaling method provides the ability to mitigate the interference.

The 802.11g standard was rapidly accepted by the market thanks to its high data rates and manufacturing cost reductions. By summer 2003, most products of mobile adapter cards and access points became dual-band/tri-mode, which means that they can work in 802.11a and 802.11b/g networks.

5. 802.11-2007 (Base Standard) [76]

In order to "roll up" a series of amendments to the 802.11-1999 version, the task group made a single specification that merged the eight amendments including 802.11a, b, d, e, g, h, i, and j with the initial standard. The merged version was approved on March 8, 2007 and became the new base standard, named as IEEE 802.11-2007.

6. 802.11n (MIMO Enhancement) [77]

The amendment, 802.11n, was published in October 2009, and improved the previous 802.11 standards by introducing multiple-input multiple-output (MIMO) antennas. It is mandatory that 802.11n devices can work in the 2.4 GHz band, and they can optionally operate in the 5 GHz bands. Modifications have been defined to both the PHY and MAC layers so that operation modes can be enabled to support a net data rate from 54 Mbps to as high as 600 Mbps. A maximum throughput above 100 Mbps can be provided as measured at the MAC data service access point (SAP).

7. 802.11-2012 (Base Standard) [79]

A single document was created to merge the ten amendments (802.11k, r, y, n, w, p, z, v, u, and s) with the 802.11-2007 base standard. The task group made much cleanup and reordered many clauses. The merged version was published on March 29, 2012, which became the new base standard and is referred to as IEEE 802.11-2012.

8. 802.11ac (High-rate in 5 GHz Band) [80]

   IEEE 802.11ac-2013 was an amendment based on 802.11n in the 5 GHz band
   and published in December 2013. Compared with 802.11n, 802.11ac employs
   several advanced new technologies. The channel bandwidth is increased to
   80 or 160 MHz from the original 40 MHz. Up to eight spatial streams can
   be supported instead of four streams. It also adopts the modulation of up to
   256-QAM, which has a higher order than the original 64-QAM. In addition,
   802.11ac introduces the multi-user MIMO (MU-MIMO). By October 2013,
   the 802.11ac products have appeared, which could support 80 MHz channels,
   256-QAM modulation, and three spatial streams in the 5 GHz band. The
   implementation yielded a data rate of up to 433.3 Mbps for each spatial stream
   and 1.300 Gbps in total. It has been expected to support four spatial streams,
   channel bandwidth of 160 MHz, and MU-MIMO in 2015, which can provide
   multi-gigabit throughput.

9. 802.11ad (High-rate in 60 GHz Band) [81]

   IEEE 802.11ad specifies a new physical layer which operates in the 60 GHz
   millimeter wave (mmWave) spectrum. The channel propagation characteristics
   in this frequency band are drastically different from the 2.4 and 5 GHz bands
   (which will be discussed in details in Sect. 2.2). The peak transmission rate of
   802.11ad is expected to achieve 7 Gbps. The WiFi Alliance is now developing
   the certification for 802.11ad.

10. 802.11af (High-rate in VHF and UHF Bands) [82]

    Approved in November 2013, IEEE 802.11af is a new amendment that specifies
    the operation of WLAN in TV white space spectrum in the VHF and UHF
    bands from 54 to 790 MHz. Hence this standard is called "White-Fi" and
    "Super Wi-Fi". The primary users in this band include analog TV channels,
    digital TV channels, and wireless microphones. In order to transmit through the
    unused channels within this spectrum, the cognitive radio technology is used
    in 802.11af to measure the interference and ensure that the interference to the
    primary users is limited. Furthermore, APs and stations may use a positioning
    method such as GPS to determine their positions and then query a geolocation
    database (GDB) through the Internet. The GDB is usually provided by a regional
    regulatory agency. According the GDB, the 802.11af devices can determine the
    available frequency channels that can be used at a given location and time.

    The modulation technique in the physical layer of 802.11af uses OFDM,
    based on 802.11ac. The advantages of working in the UHF and VHF bands are
    that, compared with the 2.4 and 5 GHz bands, the attenuation by brick, concrete,
    and other construction materials is smaller and thus the propagation path loss is
    reduced. Consequently, the coverage range may be increased.

    According to the regulatory domain, the allowed frequency band of 802.11af
    ranges from 6 to 8 MHz. Furthermore, in order to increase the channel band-
    width, at most four channels can be aggregated to form one or two contiguous
    blocks. In 802.11af, MIMO operation is supported. By using either multi-user
    (MU) or space-time block code (STBC), up to four streams are possible in
    the MIMO operation mode. Each spatial stream can achieve the data rate of

26.7 Mbps (if the channel bandwidth is 6 and 7 MHz) and 35.6 Mbps (if 8 MHz channels are used). The maximum data rates of 426.7 Mbps (6 and 7 MHz channels) and 568.9 Mbps (8 MHz channels) can be achieved by utilizing four bonded channels and four spatial streams.

11. 802.11ah (High-rate in sub 1 GHz Bands)

A WLAN working at sub 1 GHz license-exempt bands is specified in IEEE 802.11ah which is planned to be approved in 2016. Different from the conventional 802.11 WLANs operating in the 2.4 and 5 GHz bands, 802.11ah can achieve a wider coverage range thanks to the favorable propagation characteristics (such as smaller path loss) in the low frequency spectrum. Because the available bandwidth is relatively narrow, the applications can be supported by 802.11ah include range-extended hotspots, large-scale sensor networks, and outdoor WiFi to offload the cellular traffic.

12. 802.11ax (High Efficiency WLANs)

The IEEE 802.11ax amendment was initiated by the high efficiency WLANs (HEW) task group in 2014, and is expected to be released in 2019. 802.11ax is proposed to support future dense networks where multiple WLANs co-exist and each of them may have many stations. In order to address this challenge, spatial reuse of the channel resources is adopted. Furthermore, as the successor to 802.11ac, 802.11ax further increases the network resource utilization and efficiency. It is supposed to be able to increase the throughput of 802.11ac by four times. New technologies such as MU-MIMO and OFDMA are employed in both downlink and uplink to improve the network throughput. To be backward compatible with the current 802.11 WLANs, the legacy PHY preamble is adopted and EDCF is the basic MAC protocol.

As an amendment to the 802.11 standard, IEEE 802.11ai plans to introduce new techniques to setup initial links quickly. In order to use WLANs in the 45 GHz spectrum which is unlicensed in China and some other regions, the standard 802.11ad is rebanded in this band in the amendment of 802.11aj. The amendment 802.11u [78] defines the mechanisms for device discovery, and it is extended in 802.11aq that can perform the pre-association discovery of services. By using 802.11aq, a device may discover the services running on another device or provided by a network. On the other hand, for the support of multimedia services, the amendment 802.11e [54] defines the MAC procedures to support applications with QoS requirements, including the transport of voice, audio, and video over IEEE 802.11 WLANs.

(3) Medium Access Mechanisms

The contents of IEEE 802.11 standard mainly include the specification of the physical layer and the MAC sublayer. The basic 802.11 MAC sublayer defines two mechanisms for channel access: the *distributed coordination function (DCF)* and *point coordination function (PCF)*. The hybrid coordination function (HCF) is further defined in 802.11e. The MAC architecture can be illustrated in Fig. 2.1, where the functionalities of PCF and HCF are provided through the services of DCF.

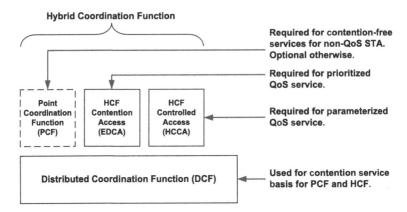

Fig. 2.1 MAC architecture defined in IEEE 802.11e

(a) IEEE 802.11 DCF

DCF is the fundamental protocol for the 802.11 WLANs [65]. It is mandatary and the basic medium access mechanism for both ad hoc and infrastructure modes. It employs the carrier sense multiple access with collision avoidance (CSMA/CA) algorithm. In DCF, stations contend for the transmission opportunities (TXOPs) in a distributed manner. A TXOP is a time interval when a particular station has the right to initiate transmissions. Using a backoff mechanism, each station should wait for a random backoff interval before transmitting or retransmitting a frame for collision avoidance. DCF is well known to provide best-effort service for asynchronous data transmission. However, since the application QoS requirements are not considered, traffic classes are not differentiated and the channel access priority is not supported. The description, analytical model, and performance evaluation of the basic DCF protocol will be discussed in details in Chap. 3.

Different from the wired networks such as Ethernet, wireless networks may suffer from the hidden terminal problem. In a WLAN with an AP, the stations communicating with the AP may not hear each other (such as the stations on the opposite edge of the geographical coverage area of the AP). This is because the wireless signal may attenuate too much before it can reach that far. In order to avoid the hidden terminal problem, DCF introduces a virtual carrier sense mechanism which is optional for a WiFi device. In this mechanism, the source and destination stations exchange short request-to-send (RTS) and clear-to-send (CTS) frames with the network allocation vector (NAV) to block the neighboring stations including the hidden terminals from transmitting simultaneously. The RTS/CTS scheme can solve the hidden terminal problem, but at the same time it introduces the exposed terminal problem, i.e., it may unnecessarily block other transmissions that will not interfere with the ongoing one.

(b) IEEE 802.11 PCF

PCF is an optional MAC technique defined in the IEEE 802.11 standards by which the channel access control is centralized. PCF adopts a centrally controlled polling method and an AP in a WLAN serves as the point coordinator to coordinate the frame transactions within the network. Therefore, with PCF, the network is operated in the "infrastructure" mode. PCF is performed based on DCF in the 802.11 MAC sublayer architecture. An AP waits for a PCF interframe space (PIFS) duration instead of the DCF interframe space (DIFS) duration to access and occupy the channel. Since PIFS is shorter than DIFS, the AP is granted a higher priority to access the wireless channel. The AP sends contention-free-poll (CF-Poll) frames to the stations that are able to operate in the PCF mode, and permits one of them to transmit a frame.

Due to the priority of PCF over DCF in channel access, the DCF-only stations might not gain access to the medium. To ensure that all stations have transmission opportunities, alternative intervals have been designed to provide both (contention-free) PCF access and (contention-based) DCF access. The AP sends beacon frames with fixed intervals, for example every 0.1 s. The time between two beacon frames is divided into two periods, the contention free period (CFP) and the contention period (CP), which are repeated continuously. When stations hear a beacon frame, they start their NAVs for the CFP period to halt the channel access. During the CFP time, the AP sends CF-Poll frames to all of the stations. It sends one frame to a station at a time in order to provide it an opportunity to send a frame. Then during the CP, DCF is used and stations can contend for the channel.

PCF allows for a better management of QoS by the centralized control. However, although PCF is more suitable to support synchronous data transmissions in a WLAN, classes and priorities of traffic which are usually adopted in other QoS mechanisms are not defined. PCF is optional and not required in the interoperability standard by the WiFi Alliance. Consequently, it is rarely implemented in the 802.11 network interface cards in practice.

(c) IEEE 802.11e EDCA in HCF

As discussed in Chap. 1, multimedia applications such as video and audio have QoS requirements such as bandwidth, delay, jitter, and packet loss which are different from data service. Wireless multimedia extensions (WME) is an interoperability certification of the WiFi Alliance, and also known as WiFi multimedia (WMM). WME proposed the IEEE 802.11e-2005 standard [54] which defines the fundamental QoS support in an IEEE 802.11 network. The 802.11e amendment specifies a set of modifications to the MAC layer to enhance QoS provisioning.

A new coordination function, named *hybrid coordination function (HCF)*, has been added in 802.11e to enhance the MAC protocols of DCF and PCF. HCF is similar to the mechanisms specified in the legacy 802.11 MAC and has defined the contention-based and reserved contention-free channel access schemes: *enhanced distributed channel access (EDCA)* and *HCF controlled channel access (HCCA)*, respectively. In order to support QoS provisioning in 802.11 WLANs, HCF introduces a number of QoS-oriented mechanisms and frame subtypes.

Four access categories (ACs) of traffic are prioritized by WME, which are voice, video, best effort, and background. Table 2.3 lists the mapping from the eight user priorities (UPs) defined in 802.1D to the four ACs. The primary principle to provide QoS in EDCA is to give the multimedia traffic a high priority and the best-effort data traffic a low priority in channel access. For example, emails are assigned with a low priority (best effort), and voice over wireless LAN (VoWLAN) and streaming videos are usually assigned to a high priority (voice and video). The channel access of each AC follows DCF but uses a set of differentiated EDCF channel access parameters. If a frame from a higher-priority flow is to be sent, it waits for less time on average than that with a lower priority. As a result, the traffic with a higher priority has a better chance of accessing the channel and being sent. This is accomplished through the modification of the backoff parameters in the traditional CSMA/CA. Thus, delay-sensitive data are protected and QoS is better supported.

Figure 2.2 illustrates the parallel backoff entities in a single IEEE 802.11e station. The traffic flows belonging to the four ACs are handled by four independent backoff entities, and an arbitration is performed inside a station to handle the internal collision among the entities.

In EDCA, a station can access the channel without contention during the period of its TXOP. Within the bounded time period of a TXOP, a station can send a number of frames given that the transmissions do not exceed the duration limit of the TXOP. In the case that a frame is too large to be transmitted within a single TXOP, the station should fragment the frame into multiple frames with a smaller size. By using the time limit of TXOPs, it can be avoided that a low-rate station occupies too much channel time to transmit frames in an 802.11 WLAN.

In addition, block acknowledgment (B-ACK) is adopted which can acknowledge an entire TXOP by using a single ACK frame. This scheme can reduce the overhead of the acknowledgment especially when TXOPs are long and multiple frames are delivered within one TXOP. Furthermore, in supporting QoS, the class of service is defined with two values: QoSAck and QoSNoAck. QoSNoAck is used to inform that a frame is not acknowledged. Thus, retransmissions of highly time-critical data (such as real-time VoWLAN), which are unnecessary, can be avoided.

The released IEEE 802.11-2007 standard has included this amendment to provide statistical instead of hard QoS support. The channel access mechanism of the EDCA protocol will be analyzed and evaluated in details in Chap. 3.

(d) IEEE 802.11e HCCA in HCF

HCCA is similar to PCF. However, there are several critical differences between them as listed below.

- In PCF, the time duration between two adjacent beacon frames is partitioned into two intervals, CFP and CP. In HCCA, CFPs are allowed to start at anytime inside a CP. Such CFP is referred to as the controlled access phase (CAP) in 802.11e. An AP can initiate a CAP any time when it wants to transmit a frame to or receive a frame from a station by contention-free channel access. During a CAP, the access to the wireless channel is controlled by the hybrid coordinator (HC), i.e., the AP. On the other hand, inside a CP, all stations contend for the channel access via EDCA.

Table 2.3 Priority to access category mappings

Priority	User Priority (UP)	AC Index (ACI)	Access Category (AC)	Designation (informative)	Minimum CW Size	Maximum CW Size	AIFSN
Lowest $\rightarrow$ Highest	1	0	AC_BK	Background	aCW_{min}	aCW_{max}	7
	2	0	AC_BK	Background			
	0	1	AC_BE	Best Effort	aCW_{min}	aCW_{max}	3
	3	1	AC_BE	Best Effort			
	4	2	AC_VI	Video	$\frac{aCW_{min}+1}{2} - 1$	aCW_{min}	2
	5	2	AC_VI	Video			
	6	3	AC_VO	Voice	$\frac{aCW_{min}+1}{4} - 1$	$\frac{aCW_{min}+1}{2} - 1$	2
	7	3	AC_VO	Voice			

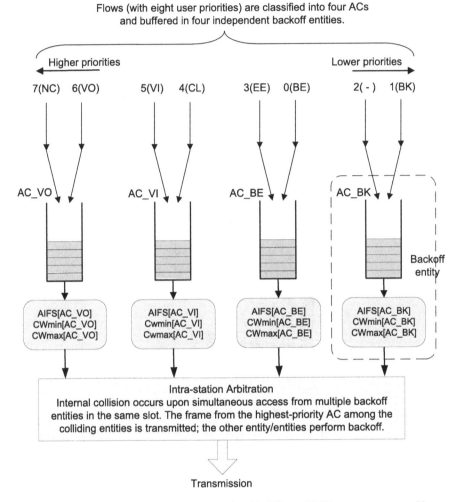

Fig. 2.2 Four parallel backoff entities for the ACs with different EDCA parameter sets and intra-contention in one IEEE 802.11e station

- As mentioned earlier, PCF does not define prioritized classes. HCCA specifies ACs and traffic streams (TS). Thus an HC can construct a queue for each session (stream), rather than for each station. In addition, HCF can coordinate these streams or sessions in any fashion (not just round-robin). Meanwhile, the stations report their queue lengths of all ACs to the HC, and then the HC may adjust the scheduling accordingly.
- In HCCA, a station may send multiple frames in a burst inside a given period of time determined by an HC, while this mechanism is not provided in PCF.
- During a CAP in HCCA, stations can also send CF-Poll frames to the HC to request data transmissions.

It is considered that HCCA is the most advanced (and complex) coordination function in the IEEE 802.11 standard family. In HCCA, QoS for each individual traffic flow can be specified with high accuracy by specific transmission parameters (such as data rate, jitter, etc.). Therefore, multimedia applications can be supported more effectively in a WLAN. However, since HCCA functionality is not mandatory in 802.11e, there are few (if any) AP products that implement HCCA at present.

2.2 Wireless Personal Area Networks

(1) Overview of WPANs

A personal area network (PAN) is used for the interconnection of devices inside the space of an individual person (intra-personal communications), sometimes with connection to a wide area network such as the Internet or cellular networks. The functionality of a PAN is to realize the inter-operation and data exchange among devices and systems at home or in a business workplace, for data and multimedia services. In addition to consumer services, WPAN applications include industrial automation and control, medical monitoring, etc. "Plug-in" is a key principle in wireless PAN (WPAN) technology. The objective is that, when two or more WPAN devices are inside the proximity of each other, they are able to communicate as if they were connected by cables. The proximity of a device is usually within several meters. Therefore, WPAN is regarded as a short-range networking technology.

WPANs are mainly standardized by the 802.15 working group of the IEEE 802 standardization committee. The 802.15 working group includes seven task groups (TGs), and they have worked out a series of specifications for the lower network layers (PHY and MAC) for various communication technologies and applications. The standards are briefly listed below, and the important ones, such as low-rate WPANs based on Bluetooth and high-rate WPANs based on ultra wideband and millimeter-wave technologies, will be discussed in details later in this section.

- TG 1: *Bluetooth*. It defines the WPAN specifications based on the Bluetooth technology for wireless connection among fixed and portable devices. The standards were approved in 2002 [83] and 2005 [84].
- TG 2: *Coexistence*. Released in 2003, the IEEE 802.15.2-2003 specification [85] addressed the coexistence issue of WPANs and other wireless networks, such as WiFi, which also operate in the same unlicensed frequency bands.
- TG 3: *High Rate WPAN*. The standard IEEE 802.15.3 defines the high-rate (11–55 Mbps) WPANs, and has three subdivisions of 802.15.3 [86] (usually called 3a), 3b [87], and 3c [88]. They employ the ultra-wideband (UWB)-based and mmWave-based PHYs.
- TG 4: *Low Rate WPAN*. The standard IEEE 802.15.4-2003 [89] specifies the WPAN that operates at low data rate, has very long battery life (e.g., for months or even years), and has low system complexity. The 1st edition was released in May 2003. A number of standardized protocols and proprietary protocols have been

developed to work on top of the 802.15.4-based wireless networks, including IEEE 802.15.4-2015 [97], ZigBee, IPv6 over low power wireless personal area networks (6LoWPAN), wireless highway addressable remote transducer protocol (WirelessHART), and international society of automation (ISA) 100.11a. The low-rate WPAN will be discussed in details in the next section.

- TG 5: *Mesh Networking*. IEEE 802.15.5 [98] defines the architecture for WPAN devices to form an inter-operable, steady, and flexible wireless mesh network. The standard consists of two divisions: mesh-less WPAN with low data rate and mesh WPAN with high data rate. The former is established on the MAC mechanism of IEEE 802.15.4-2006 [90], while the latter is based on IEEE 802.15.3b MAC [87].

- TG 6: *Body Area Networks*. IEEE 802.15.6 TG approved the standard for body area network (BAN) technologies in December 2011 [99]. It is a wireless standard for low-power and short-range devices which operate on, in, or around an individual body. The wireless BAN (WBAN) enables a spectrum of applications including medical care, consumer electronics, and personal entertainment. The BAN will be discussed in details in the next section.

- TG 7: *Visible Light Communication*. The IEEE 802.15.7 [100] draft was completed in December 2011 for visible light communications (VLC). It is especially used for optical communications using visible light in free space.

(2) Low-Rate WPAN: IEEE 802.15.4

(a) Overview and Physical Characteristics

IEEE 802.15.4 defines the PHY scheme and MAC protocol for low-rate WPANs (LR-WPANs). The MAC protocols in the standards, including ZigBee, ISA100.11a, WirelessHART, and MiWi (designed by Microchip Technology), are based on 802.15.4 and each of them further defines different upper layers such as packet routing and applications. For example, 6LoWPAN utilizes 802.15.4 (for lower layers) and the standard Internet protocols (for higher layers) to build a wireless embedded Internet. The key feature of 802.15.4 is to provide low-rate, low-cost, short-range communications among nearby devices, contrasted to the high-rate, more power-intensive wireless solutions such as WiFi. The target capacity of 802.15.4 is a link of 10 m at a transmission rate of only 250 Kbps or even lower (such as 20, 40, and 100 Kbps) to achieve extremely low power consumption. It also emphasizes the need for little to no underlying infrastructure for low operation costs and easy and flexible network setup.

In IEEE 802.15.4, PHY is composed of the physical signal transceiver in the radio frequency (RF) band, and also selects appropriate channel for networking and managing energy consumption. The original version of the standard released in 2003, IEEE 802.15.4-2003 [89], specifies the PHY utilizing the direct sequence spread spectrum (DSSS) technique which operates in three unlicensed frequency bands.

- 868.0–868.6 MHz: Europe, one communication channel, transfer rates of 20 and 40 Kbps;

- 902–928 MHz: North America, up to ten channels (2003) and extended to thirty (2006), transfer rates of 20 and 40 Kbps;
- 2400–2483.5 MHz: worldwide use, up to sixteen channels, transfer rates of 250 Kbps.

The 2006 revision, IEEE 802.15.4-2006 [90], improves the maximum data rates up to 100 and 250 Kbps in the frequency bands of 868 and 915 MHz. A series of amendments following the original versions are briefly listed below.

- *IEEE 802.15.4a-2007: WPAN Low Rate Alternative PHY [91]*. This amendment specifies two additional PHYs, one using direct sequence (pulse radio) UWB (operating in the unlicensed UWB spectrum, including below 1, 3–5, and 6–10 GHz) and the other using chirp spread spectrum (operating in the unlicensed 2450 MHz spectrum). The radio pulse-based PHY schemes are able to perform localization and ranging with high precision (e.g., the accuracy can be one meter or even smaller), large aggregate throughput, and scalability in the tradeoff between longer range and higher data rates. They also provide lower power consumption options with reduced cost.
- *IEEE 802.15.4-2006: Revision and Enhancement [90]*. Approved in June 2006 and published in September 2006, 802.15.4-2006 specifies enhancements and clarifications to the IEEE 802.15.4-2003 standard. The enhancements include resolving ambiguities, reducing complexity if not needed, increasing scalability in the use of security key, considering frequency bands that are newly available, and others.
- *IEEE 802.15.4c: PHY Amendment for China [92]*. It was approved in 2008 and published in January 2009. It adds the newly opened RF spectrum bands in China for WPAN use, including the 314–316, 430–434, and 779–787 MHz bands.
- *IEEE 802.15.4d: PHY and MAC Amendment for Japan [93]*. It defines a new PHY and the necessary MAC modifications to operate in the newly allocated frequency bands from 950 to 956 MHz in Japan. Meanwhile, it ensures the coexistence between 802.15.4 WPANs and passive tag systems in this band.
- *IEEE 802.15.4e: MAC Amendment for Industrial Applications [95]*. It was approved in February, 2012. As the industrial markets increase, 802.15.4e defines a new amendment to the MAC scheme in the existing standard 802.15.4-2006. Channel hopping strategy is employed to improve the signal robustness against external interference and continual multipath fading.
- *IEEE 802.15.4f: PHY and MAC Amendment for Active RFID [94]*. It defines new wireless PHY and improvement to the 802.15.4-2006 MAC to enable the localization applications of active RFID systems.
- *IEEE 802.15.4g: PHY Amendment for Smart Utility Networks (SUN) [96]*. Released in April 2012, the 802.15.4g standard creates a PHY amendment that is capable of supporting large, geographically diverse networks. This standard can facilitate very large-scale process control. For example, the utility smart grid network may have millions of fixed terminals with minimal infrastructure.

(b) MAC and Network Topology

The MAC layer manages the physical channel access for frame transmissions, network beaconing, time slot scheduling, frame validation, and node associations. Note that the majority of the IEEE 802.15.4 PHYs only allow a frame size of up to 127 bytes and do not provide the exchange of standard Ethernet frames to upper layers. Therefore, the adaptation layer (like in 6LoWPAN) is needed to provide frame fragmentation for network-layer packets in the protocol stack. For the channel access mode, 802.15.4 includes both guaranteed time slots by reservation and transmission opportunities by contention with collision avoidance (following the CSMA/CA mechanism).

Network nodes in the 802.15.4 standard are defined as two types. The first kind is the full-function device (FFD) which operates as the coordinator of a PAN as well as a common node. An FFD node can also forward/relay packets. On the contrary, the reduced-function devices (RFDs) have much less complexity. They typically have limited resource and communication capacity, and can talk with FFDs only. An RFD cannot operate as a coordinator.

The 802.15.4 WPAN can work in either the peer-to-peer mode or the star mode. An 802.15.4 network requires at least one FFD to serve as the coordinator which is in charge of the establishment of the whole network. Every device is assigned a unique identifier. Within each PAN domain, short 16-bit identifiers may be used. Peer-to-peer networks can work in the ad hoc mode and are able to perform self-organization and management. Furthermore, the network can be extended to become a generic mesh network. The network nodes are organized in clusters. There is a local coordinator (an FFD node) in every cluster and a global coordinator in charge of the whole network. The cluster-based network topology is illustrated in Fig. 2.3. The cluster-tree topology of ZigBee supports power-saving operations and light-weight tree routing protocol, and hence is suitable for low-power low-rate WSNs. However, events in the sensing area may trigger sensors to generate much higher data traffic. Due to the restricted routing protocol, the cluster-tree networks may not be able to deliver the increased traffic. Huang et al. [52] have proposed an adoptive-parent based framework to increase the bandwidth utilization. The throughput optimization is formulated as a convex-constraint maximum flow, and a distributed pull-push-relabel (PPR) algorithm compatible with the existing standard is designed to maximize the throughput. On the other hand, the 802.15.4 standard also supports the star topology which is more structured. In this case, the coordinator of a star-like network is required to be the central node.

Frames are the basic unit for data transfer in 802.15.4. There are four fundamental types of frames, which are data, acknowledgment, beacon, and MAC command. Based on the definition of the four frame types, a reasonable tradeoff can be made between network simplicity and robustness. The superframe structure which is limited with two beacons sent by the coordinator is used for node synchronization and channel access management. In peer-to-peer networks, the communications between any two devices are possible. The channel access can be coordinated in

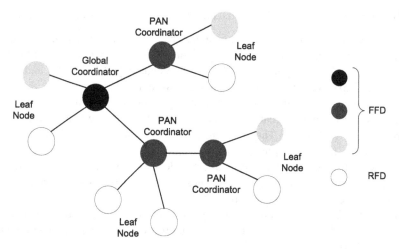

Fig. 2.3 IEEE 802.15.4 cluster-tree based network topology

two ways, by either unslotted CSMA/CA based on contention or synchronization mechanisms. However, in the star topology, nodes cannot communicate with each other except with the network coordinator.

(3) Low-Rate WPAN: Bluetooth

(a) Overview and Physical Characteristics

Bluetooth uses short-range radio communications for exchanging data over approximately 10 m from fixed and mobile devices and building WPANs. Bluetooth is managed by the Bluetooth special interest group (SIG), which was formally created on May 20th, 1998, and now has more than 20,000 member companies. Although Bluetooth used to be specified as IEEE 802.15.1, it is not maintained any more by IEEE 802 group but by the SIG. Bluetooth is defined for low-rate, short-range, low-power, and low-cost wire-replacement communications. Moreover, by simplifying the discovery and setup of services between devices, Bluetooth operates well with simple setup in the scenario that two devices can be connected with minimum configuration, contrasted to the WiFi technology which requires client configuration and provides high speeds. The WPAN devices using Bluetooth including keyboards, pointing devices, audio head-sets, laptops, printers, smart phones, and game consoles have become very popular today.

Bluetooth works in the spectrum band of 2400–2483.5 MHz, the globally unlicensed ISM band. The range is divided into 79 designated channels, and the frequency-hopping spread spectrum (FHSS) scheme is adopted in the PHY layer. The original data stream is partitioned into packets and every packet is sent in one channel. The bandwidth of a channel is 1 MHz (note that 2 MHz spacing is used in Bluetooth 4.0, resulting in 40 available channels). When adaptive frequency-hopping (AFH) is enabled, PHY usually executes 1600 hops in 1 s.

The development of the Bluetooth technology is briefly reviewed below.

1. Bluetooth v1.0 and v1.0B contain a number of issues, which led to much difficulty in the inter-operation among products from different manufacturers.
2. Bluetooth v1.1 was approved as the IEEE Standard 802.15.1-2002 and many problems existing in the v1.0B standard were fixed. In addition, the non-encrypted channels and received signal strength indicator (RSSI) were added.
3. Bluetooth v1.2 made great improvement. The fast discovery and connection mechanism was added. The AFH in the FHSS, which excluded the crowded frequencies in the hopping frequency sequence, was introduced. Thus the resistance to radio interference from other devices in the unlicensed band was enhanced. Higher transmission data rates up to 721 Kbps were provided. v1.2 was released as the IEEE standard 802.15.1-2005.
4. Bluetooth v2.0+EDR was released in 2004, which defined an optional enhanced data rate (EDR) up to 3 Mbps for faster data transfer (the practical data transfer rate is 2.1 Mbps).
5. Bluetooth v2.1+EDR was accepted by the SIG on July 26th, 2007, which provides secure simple pairing (SSP) to achieve enhanced security.
6. Bluetooth v3.0+HS was approved by the SIG on April 21st, 2009. The key enhancement was to support high-speed (HS) transport by employing 802.11, which was an alternative MAC/PHY (AMP) for Bluetooth. The link negotiation and establishment were done by Bluetooth communications, while a collocated 802.11 link was used to carry high-rate data traffic. Thus, the data rate could theoretically achieve 24 Mbps.
7. Bluetooth v4.0 [15], also named as Bluetooth Smart, was approved on June 30th, 2010. The Classic Bluetooth (legacy Bluetooth protocols), Bluetooth HS (based on WiFi), and Bluetooth low energy protocols are included in this release.
8. Bluetooth v4.1 [16] was formally released by the SIG on December 4th, 2013. The update adds new features which improve co-existence support for LTE, transfer speed for bulk data, and simultaneous multiple role support of a device which can be used for developer innovation.

(b) MAC and Network Topology

Bluetooth is a packet-based protocol with a master-slave structure and one master may communicate with up to seven slaves. Therefore, the eight devices including the master form a PAN called a *piconet* (the prefix "pico" means very small or one trillionth). The range of a piconet is typically 10 m. In a piconet, the first Bluetooth device plays the role as the master, and then other devices which communicate with the master are slaves. All devices in a piconet use the common clock of the master. The master typically selects a slave device to interact with and it switches rapidly among the devices in a round-robin fashion.

Furthermore, as specified in the Bluetooth standards, two or more piconets can inter-connect with each other and form a scatternet. In this case, a device may need to be the master in one piconet and at the same time a slave in another piconet, in order to realize the inter-connection of multiple piconets.

(4) High-Rate WPAN: 3.1–10.6 GHz UWB

(a) Overview and Physical Characteristics

The frequency band from 3.1 to 10.6 GHz has been issued by FCC for the unlicensed use of UWB devices [40]. In addition, the transmission power must be below the power spectrum mask (−41.25 dBm/MHz for indoor deployment), in order to allow physically coexistence with other 802.15 devices which also operate in this band. For the purpose of communications, a UWB system is defined in either one of the two ways. The first way is that the −10 dB fractional bandwidth of the communication system should be at least 20 %. The second is that −10 dB bandwidth needs to be equal to or more than 500 MHz. As a promising technology for short-range, high-rate, and low-power wireless communications, UWB technology has received significant interests from both academia and industry [118, 162]. Due to the very high-data rate over 100 Mbps, UWB networks are able to support simultaneously several isochronous streaming flows, such as voice over IP (VoIP), high-definition television (HDTV), and massive data download [35].

The IEEE 802.15.3 working group is responsible for specifying the PHY and MAC standards for high-rate WPAN. It has three subgroups, IEEE 802.15.3a [86], 3b [87], and 3c [88]. Since 3c specifies the UWB technology operating in the mmWave band, it will be discussed in the next subsection. The works of 3a and 3b are summarized as below.

- *IEEE 802.15.3a* provides a high speed UWB PHY enhancement amendment for applications involving streaming multimedia and massive data. There were two proposals for the UWB technologies, multi-band orthogonal frequency division multiplexing (MB-OFDM) and direct sequence UWB (DS-UWB). They were proposed by two industry alliances comprising of different companies. However, an agreement on choosing which technology proposal could not be achieved among the task group members. Consequently, the group was dissolved in January 2006. The UWB-based high-rate WPAN will be introduced in details later in this section.
- *IEEE 802.15.3b-2006* amendment was issued on May 5, 2006. The amendment enhanced the implementation and inter-operability of the MAC, which included minor optimizations, error corrections, ambiguity resolution, and editorial improvement.
- *IEEE 802.15.3c-2009* was released on September 11, 2009. The TG3c was organized in March 2005 and developed an alternative PHY for WPAN based on mmWave. This mmWave PHY works in the unlicensed frequency band from 57 to 64 GHz, which is a clear band as defined by FCC 47 CFR 15.255. Operating in this mmWave band, a very high speed over 2 Gbps is possible. The targeted applications include high-data rate Internet access, massive data download, real-time streaming (high-definition IPTV, video on demand, etc.), and cable replacement (e.g., wireless home theater).

As a very wide band radio technology, UWB transmission has several unique merits, for example, very low transmission power, capability in mitigating multipath effect, high-data rate, and ability in precise positioning. The UWB technology is expected to fully exploit the very large bandwidth in transmitting information, and thus tries to provide a data rate higher than 100 Mbps. By restricting the transmission power under the appropriate level required by FCC, UWB is able to coexist with other wireless systems and share the spectrum.

UWB was originally employed as "pulse radio" in radar for positioning. For high-rate data communications (> 100 Mbps), the UWB system can be implemented by a pulse-based approach [136, 146, 162] or an MB-OFDM based approach [7, 114]. In the former approach, information is modulated on very short pulses. The duration of a pulse is typically in the order of a nanosecond. In the latter approach, the combination of frequency hopping and OFDM is adopted. For the data transmissions in a multipath wireless channel, either of the two UWB proposals has its pros and cons.

- In the pulse-based UWB, the multipath diversity can be exploited effectively by utilizing the rich resolvable multipath components (MPCs).
- A long channel acquisition time is needed for channel estimation in the pulse-based approach. Also, high-speed analog-to-digital conversion (ADC) is required in processing the received pulse signals.
- MB-OFDM has the advantage in spectral flexibility and efficiency. It has also the robustness against narrowband interference.
- MB-OFDM needs a slightly more complicated transmitter compared with the pulse-based UWB system.

In addition, the pulse-based UWB technology is also flexible in providing low-rate data transfer (< a few Mbps) over moderate or long distances (from 100 to 300 m) [107, 111].

(b) WiMedia WPAN MAC

The function of MAC in WPANs is similar to that in WLANs, i.e., coordinating the channel access from the competing stations in order to transmit data efficiently and fairly. The major MAC specifications for UWB-based WPANs include IEEE 802.15.3 [86] and WiMedia-368 [69] which is specified by the Multiband OFDM Alliance (MBOA).

Based on the MB-OFDM technology, the WiMedia Alliance has released the standards for both the PHY scheme and the MAC protocol [49]. WiMedia UWB has been promoted for personal computers, consumer electronics, mobile devices, and automotive networks. In order to achieve ad hoc connectivity, two distributed channel access mechanisms are defined by the WiMedia specification: the prioritized channel access (PCA) and the distributed reservation protocol (DRP).

PCA is a contention-based approach and it adopts the mechanisms that are similar to EDCA in IEEE 802.11e to provide differentiated channel access. A device senses the channel before transmitting frames. To prioritize traffic channel access, the parameters in the backoff and channel contention are selected depending on the traffic class and priority. Only statistical QoS provisioning is realized in the EDCA-like MAC protocols. Therefore, the QoS requirements of isochronous traffic, such as the stringent delay bound, are difficult to be satisfied in PCA [21]. The detailed performance analysis on the PCA protocol can be found in [63] and [138]. The analysis and performance evaluation of EDCA will be presented in details in Chap. 3, which is readily applicable to PCA.

DRP is a distributed TDMA protocol, by which users reserve channel access time in superframes and then transmit frames within the reserved time slots without interruption. However, different from the original TDMA scheme, the stations in a WPAN first negotiate the channel reservation in a distributed manner. The advantage of DRP is that the transmission opportunities and time for isochronous traffic are guaranteed by the channel reservation. Since the QoS requirements such as the stringent delay bound is satisfied, DRP is preferable for streaming multimedia applications. However, isochronous data flows such as compressed video traffic have bursty data rates. The packet inter-arrival time is random (i.e., the instantaneous data rate varies significantly). The difference between the reserved bandwidth and the time-varying requirement from the traffic results in the difficulty in resource reservation. To accommodate the burstiness of the traffic flow, over-reservation is usually adopted which leads to considerably inefficient utilization of the network bandwidth when fixed channel time is reserved (*hard reservation*). This inefficiency in resource reservation can be improved via the *soft reservation*. In the latter, the unused reserved time can be released by the owner and other stations which have backlogged frames can contend for channel access following the PCA protocol.

Another important aspect of DRP is that the allocation of the reserved time slots for one flow is more flexible. The main feature of DRP different from the traditional TDMA scheme is in the reservation pattern. To limit the delay variation, it is desirable to reserve contiguous or evenly spaced time slots with constant interval in each scheduling cycle (i.e., a superframe in the MAC protocol). Such reservation pattern can be realized by a centralized coordinator. However, in DRP, because the locations of the available time slots within a superframe are arbitrary, there can be multiple reserved time slots which may be non-uniformly distributed inside a superframe for one traffic flow, resulting in a random reservation pattern in a scheduling cycle. The reservation algorithms for multimedia traffic in DRP will be discussed in details in Chap. 4.

(5) High-Rate WPAN: 60 GHz PHY

(a) Overview and Physical Characteristics

With the hope that all cables in home networks for indoor information delivery are replaced by high-rate wireless data bus, FCC issued the frequency band from 57–64 GHz that became available in FCC 95-499 [3] and CFR 15.255. Japan declared the 59–66 GHz band and the European Telecommunications Standards Institute (ETSI) allocated the 57–66 GHz band too. Thus, a common, contiguous 5 GHz band is available around 60 GHz in major markets around the world [9] (see Fig. 2.4). Since the signal wavelength at 60 GHz is about 5 mm, this spectrum is referred to as the mmWave band. In July 2003, with the increasing interests in developing an mmWave PHY within the IEEE 802 family, an interest group belonging to the

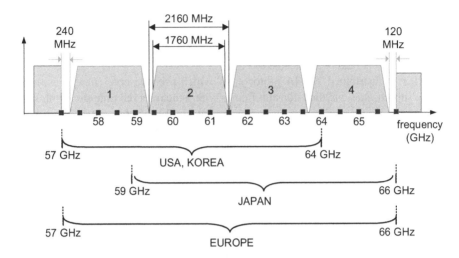

Fig. 2.4 Channelization of 802.15.3c and unlicensed bands around the globe

802.15 working group was formed for WPANs and a study group was formally established in March 2004. The members of this group developed an mmWave-based alternative PHY to support 1 Gbps or higher data rates for the existing 802.15.3 WPAN standard. It was decided by the IEEE 802 Working Group to adopt the existing MAC protocol (IEEE 802.15.3b), and necessary modifications and extensions should be done to improve implementation and interoperability of the MAC. The task group started by concentrating on the application models, an indoor wireless channel model at 60 GHz, and the evaluation criteria for PHY proposals. After 2 years of hard work, three PHY schemes and a number of MAC enhancements were proposed to realize different usage scenarios. Finally, the IEEE 802.15.3c-2009 [88] was released in September 2009, approved by the IEEE-SA standards board.

The mmWave WPAN operating in 60 GHz band allows the physical coexistence with other microwave devices communicating within this band. It also allows high data rate over 2 Gbps to support applications including massive content transfer, high-speed Internet access, real-time streaming, and wireless cable replacement. By analyzing in details the potential applications by consumers, the 802.15.3c Task Group defined five usage models (UMs) in the 60 GHz band [115].

- UM 1) *Uncompressed video streaming:* The very wide bandwidth enables sending high-definition television (HDTV) traffic flow from HD video cameras to display screens to replace video cables. The HD video signals have at least 1920×1080 pixel resolution, 24 bits for each pixel and a rate of 60 frames per second. Hence the uncompressed video streaming bit rate is more than 3.5 Gbps.
- UM 2) *Uncompressed multivideo streaming:* A home network gateway may deliver several video signals to multiple TV sets or a TV can show a couple

of channels side by side on a screen. One stream may require 720×480 pixels per frame. Thus the gateway should be able to provide at least two streams with 0.62 Gbps each simultaneously.

- UM 3) *Office desktop:* This UM considers the data communication between a personal computer and multiple external peripherals, such as one or more screens, printers, and hard disks. The data flows can be unidirectional or bidirectional. For reliable data delivery, retransmissions may be needed.
- UM 4) *Conference ad hoc operation:* Many computers are communicating with each other in an ad hoc, bi-directional, and asynchronous mode. The conference operation range is usually larger than the office desktop.
- UM 5) *Kiosk file downloading:* The portable devices will be equipped with transceivers with low complexity and power consumption to enable large data uploads and downloads. For example, downloading video files and large amount of pictures from smart phones at 1.5 Gbps within 1 m range will be possible.

(b) 60 GHz Channel Model

Different from the radio systems within unlicensed ISM bands such as 2.4 and 5 GHz, 60 GHz signals have much smaller wavelength, resulting in significantly higher propagation loss. Hence the efficient PHY technologies and MAC protocols adapting to the new features are needed. For example, directional transmission and reception are usually needed to increase the signal power in the target direction. The conventional Saleh-Valenzuela (S-V) channel model [116], which is suitable for the signal transmission/reception by the IEEE 802.11 and IEEE 802.15 specifications in the ISM band, cannot be applied for the 60 GHz propagation. Therefore, a new channel model has been proposed by the IEEE 802.15.3c channel modeling subcommittee. It adopts the two-path model which incorporates a line-of-sight (LOS) component and NLOS reflective clusters similar to the S-V model [56, 125].

The new model is illustrated in Fig. 2.5. In order to cover all the possible scenarios, the model is determined by a set of different parameters (e.g., path loss coefficients, shadowing effect [151], etc.). These parameters have been extracted from field measurements for different scenarios.

(c) MAC in IEEE 802.15.3c

The IEEE 802.15.3c standard [88] defines the PHY and MAC specifications for mmWave-based high-rate WPANs. In the standard, a group of devices (DEVs) exchange data in an ad hoc fashion and form a *piconet*. In order to ensure piconet synchronization and manage the channel access among DEVs, one DEV would play the role as the *piconet coordinator (PNC)*. Time is divided into sequential *superframes* and each superframe contains three segments: a beacon period, a contention access period (CAP), and a channel time allocation period (CTAP). Beacons are broadcast periodically by the PNC which bear the necessary control information, for example, the time and opportunities a DEV can access the channel. When a DEV hears a beacon, it would know the presence of a piconet.

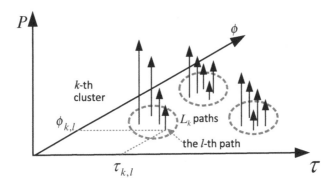

Fig. 2.5 Graphical representation of the 60 GHz channel model

The CAP is mainly used for the PNC and DEVs to exchange command and control messages. Because the packet-based data transfer is mostly asynchronous, the contention-based access scheme is adopted, i.e., the prioritized CSMA/CA such as IEEE 802.11e. The analytical model and performance analysis of CAP with the existence of the CTAP within a superframe will be presented in details in Chap. 5.

The rest of the superframe includes the CTAP which adopts reservation-based channel access such as TDMA. The CTAP comprises multiple channel time allocations (CTAs). The CTAs are time slots granted by the PNC and each CTA is used by a certain pair of DEVs for data exchange. By TDMA, time-sensitive applications such as multimedia streaming can utilize CTAs for guaranteed data delivery and the delay can be bounded.

Based on the fundamental architecture of the MAC protocol for WPANs, the task group has also developed enhancements in three major areas to define an efficient and well-structured MAC layer [9].

- *Coexistence among 802.15.3c PHYs:* For the purpose to make devices using various PHY schemes physically coexist with each other, sync frames are employed. It is mandatory that a PNC-capable DEV should transmit a sync frame in each superframe as specified by the 802.15.3c rules. A PNC is also capable of receiving and decoding sync frames regardless operating with what kind of PHY scheme. Consequently, a PNC can obtain the information about the existence of a nearby piconet. Then it has the chance to join it instead of establishing a new, independent one. Hence the sync frame mechanism is an effective way to create, maintain, and manage the coexistence of piconets and to avoid co-channel interference between nearby piconets.

- *Frame Aggregation:* In WLAN and WPAN systems, the transmission rate of the frame header is usually fixed at the lowest mandatory rate in the PHY for reliable reception. Thus, with the increase of the transmission rate of the data payload, the network efficiency will decrease because the ratio between the overhead time and the payload time increases, especially in high-speed networks. To improve transmission efficiency and effective throughput, frame aggregation is used. The principle is to concatenate the payload data from multiple MAC service data units (MSDUs) and remove the extra overhead (such as the preamble and PHY/MAC header). In the 802.15.3c standard, the standard aggregation and low-latency aggregation methods are specified.

- *Beamforming:* The high-speed WPAN is expected to achieve MAC throughput of a few gigabits per second over a short to moderate range. To accomplish this, a high received SNR is critical. In order to compensate the high propagation loss (especially for the 60 GHz band) and mitigate the attenuation caused by shadowing, directional transmission is preferable. By concentrating the signal power in the target direction, the received SNR can be increased significantly compared with the omni-directional emission. Using multiple antennas (antenna array) and beamforming is an effective way to realize directional transmission. Integrating multiple antennas into a portable device has also become feasible, because the dimension and the necessary spacing between the antennas operating at 60 GHz are in the order of millimeters [51].

In summary, as the first IEEE wireless standard for data rate higher that 1 Gbps at the MAC service access point (SAP), IEEE 802.15.3c is designed to not only develop three new PHY schemes, but also enhance the existing 802.15.3 MAC by specifying piconet management mechanism, frame aggregation, and beamforming capability. Benefitting from the spectrum regulation and standardization effort, the rapid deployment of WPANs throughout the world has become possible. The commercial products following the 802.15.3c standard have already appeared, and consumers can have the WPANs device coexistence without worrying about the interference.

There are a lot of research opportunities for WPANs. For example, because of the high propagation loss of 60 GHz signals in indoor environments, the signal coverage may be too limited to form an expected home network. Thus, repeaters and multihop solutions for a typical WLAN deployment will be necessary. The range enhancement by employing new technologies such as advanced coding and steerable antennas is also an option. It can be predicted that the new efficient MAC protocol for high-speed WPANs will consistently play an important role in improving the network throughput, delay performance, data delivery reliability, network maintenance, and QoS provisioning.

2.3 Wireless Body Area Networks

(1) Overview of WBANs

With the rapid growing demands of ubiquitous communications and great advances in very-low-power wireless technologies, there have been considerable interests in the development and application of wireless networks around humans. A body area network (BAN), also referred to as a wireless body area network (WBAN) or a body sensor network (BSN), is an RF-based wireless network that

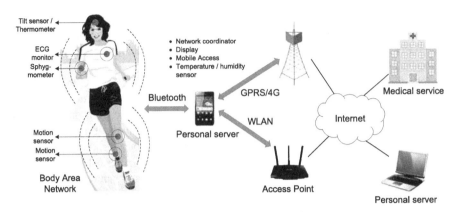

Fig. 2.6 Interconnection of WBAN, WPAN, (W)LAN, and wide area networks

interconnects tiny nodes of wearable sensor or computing devices to implement sensing and communications within the proximity of a body. With the trend towards the miniaturization of devices and wearable technology, humans can carry the BAN devices in different locations, for example, embedded inside or surface-mounted on the body in fixed positions, in pockets, by hand or in bags. BANs target diverse applications including healthcare, athletic training, workplace safety, consumer electronics, secure authentication, and safeguarding of uniformed personnel.

Based on the miniaturized devices, a BAN network usually consists of several body sensor units (BSUs) together with a single body central unit (BCU). Larger-sized communication devices (smart phones or pads) also play an important role in terms of acting as a data hub, data gateway to the Internet and/or providing a user interface to view and manage BAN functions. Typically, the transmissions of the BSUs and BCUs cover a short range of about 2 m. Through gateway devices, a BAN can be connected to local and wide area networks, as illustrated in Fig. 2.6.

WBANs will play an important role in enabling ubiquitous communications and creating a huge potential market. In the area of healthcare, according to the World Health Organization's statistics, millions of people suffer from obesity or chronic diseases every day, while the aging population is becoming a significant problem. Using the WBAN technologies, medical professionals can collect various first-hand physiological changes for health monitoring and care using the Internet regardless of the patients' locations. If an emergency is detected, the physicians can immediately inform the patient and medical service through the networks by sending appropriate messages or alarms. From the consumer electronics perspective, short-range wireless networks for human-computer interaction (HCI) and entertainment are also promising.

From the technical point of view, to design and implement such WBANs faces a series of challenges.

- The importance of reliability is obvious especially when WBANs are used for monitoring the health status of patients. Particularly any emergency signal cannot be missed.
- Energy utilization and efficiency are critical for WBAN sensors, because their battery capacities are very limited due to their tiny sizes. It is difficult to replace batteries of the implanted BSUs.
- Another important issue is inter-operability. WBAN systems need to be scalable and support seamless data delivery over various networks such as Bluetooth and ZigBee. The smooth migration through networks is required to ensure ubiquitous connectivity and coverage.
- The BAN sensors must have low complexity, small size and weight, and high power efficiency.
- Interference should be regulated. Wireless links in a WBAN should reduce the interference level to other physically close networks, and allow the coexistence especially when WBANs are densely deployed.

The application requirements such as extreme energy efficiency and the unique characteristics of the wireless channels, require novel solutions in resource management and MAC protocols of WBANs. In the following, we present some networking techniques and the studies of WBAN channels that could be used to address these challenges.

(2) The BAN Standards

Multiple standards have been proposed for WPANs, such as Bluetooth [15, 16, 117] and IEEE 802.15.4 [89, 90, 97], which are candidate approaches to address the challenges raised in WBANs. Comparisons of the traditional and new standards for the WPANS and WBANS are listed in [19]. In addition to the WPAN standards such as Bluetooth Low Energy, Bluetooth 3.0, UWB, and ZigBee, other proprietary and open technologies such as Insteon, Z-Wave, ANT, RuBee, and RFID are introduced. Insteon and Z-Wave are both proprietary specifications which define the mesh networking and can be used for home automation. Z-Wave operates in the 2.4 GHz ISM band, while Insteon utilizes both power lines and the 900 MHz ISM band. ANT is another proprietary sensor networking technology, and it features the simple protocol stack and low power consumption. ANT has been embedded in some Nike shoes to collect workout data and talk to iPod by wireless connection. RuBee and RFID are technologies to support asset management and tracking. They are complimentary specifications by utilizing different frequency bands and battery life, and used for different application scenarios.

These standards have all been implemented by application specific integrated circuit (ASIC) and are being sold in comparable volumes each year. With the advances in very large-scale integrated (VLSI) circuit, dual and even multi-standard radios are integrated on a single chip, greatly reducing the cost and hardware system complexity, which will boost the applications and deployment of WBANs.

(3) IEEE 802.15.6 MAC

The IEEE 802.15.6 working group was created in November 2007 and concentrated on developing a standard for low-power and short-range wireless networks [99]. The objective is to address the challenges in covering the wide application spectrum and tough reliability and energy efficiency requirements in WBANs. In December 2011, the 802.15.6 standard was approved for WBANs operating in the vicinity of or inside a human body (the standard is not limited to humans). The standard can support different kinds of applications as discussed earlier. The physical and MAC layer schemes are specified. Each application has specific QoS requirements, in particularly, the data rate, reliability, and energy efficiency, and these parameters may not be simultaneously optimized. One of the key features of 802.15.6 is the flexibility and manufacturers can select the proper PHY and MAC schemes for the target application scenario.

The PHY is responsible for (1) the operations of the radio transceiver, (2) channel status estimation and selection, and (3) signal processing and detection. IEEE 802.15.6 specifies three operational PHYs. The UWB and human body communication (HBC) schemes are mandatory while the narrowband one is optional. Furthermore, according to the scenario and network setting, a set of carrier frequency bands can be selected [17]. The frequency spectra which can be used by WBANs are managed by the spectrum authorities in various countries, as summarized in [129]. The technology to adopt can be selected depending on the applications and radio regulations. For example, the radio signals in lower frequency bands experience less path loss and attenuation by a human body, but UWB frequency band offers higher transmission speed. Furthermore, some particular applications are regulated to be allowed only at certain frequencies (e.g., implants may operate only at 402–405 MHz worldwide).

In the MAC layer, nodes are organized into one or two-hop star-topology wireless networks. Similar to other low-power standards, 802.15.6 utilizes a network coordinator (or hub) to control the behavior of each device. In a WBAN with two-hop topology, a node which is capable of relaying data can serve as the relay to forward frames from a node to a coordinator and vice versa.

The IEEE 802.15.6 specifies multiple channel access modes. The representative one is the beacon mode which utilizes a superframe structure for timing boundaries. The coordinator sends out *beacons* periodically to setup a timing reference. The intervals between beacons are *superframes* and each superframe is divided into small intervals called *slots* which are used by the sensor nodes in a multiple access manner. In this mode, the coordinator is responsible for allocating the slots by beacon frames. The coordinator transmits beacons in active superframes where there are transmissions inside. If there are no scheduled data to transmit, one or more inactive superframes can be inserted. As illustrated in Fig. 2.7, the superframe is divided into exclusive access phases (EAPs), random access phases (RAPs), a managed access phase (MAP), and a contention access phase (CAP). The EAPs are utilized by nodes to transfer high-priority or emergency traffic, while the RAPs and CAP are designed to accommodate bursty traffic. In the EAP, RAP, and CAP

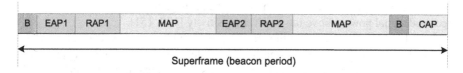

Fig. 2.7 Beacon mode with superframe boundaries in IEEE 802.15.6

periods, the coordinator may employ slotted ALOHA and CSMA/CA for UWB and narrowband PHYs, respectively. The MAP period is used for the allocation of uplink, downlink or both. Type I polled and posted allocations which are used to obtain scheduled allocations based on the access scheduling mechanism are also transferred in MAP. A detailed description of these schemes can be found in [99].

In addition, IEEE 802.15.6 introduces a polling and posting mechanism which is also referred to as "improvised and unscheduled access". In this mechanism, the coordinator informs the devices that they have been given time slots to transmit or receive data exclusively. It can select the improvised access to transmit poll or post commands. In this mode, a station may not perform pre-reservation or provide notice in advance in the beacon mode or the non-beacon mode with superframe boundaries. The commands can be employed to initiate the transactions of one or multiple data frames transmitted by the nodes or coordinator outside the scheduled allocation interval. Each device does not need to implement all of the access control mechanisms, and it can just select only those suitable for its operation.

Reliability is another key consideration in many BAN applications, especially when used for medical care. Considering that radio spectrum has been more and more crowded, the mutual interference among a variety of wireless networks is also increasing. The IEEE 802.15.6 task group has attempted to improve the robustness against interference of WBANs from various aspects. First, FCC has approved the allocation of 40 MHz of spectrum bandwidth for low-power medical BAN (MBAN) in the band of 2360–2400 MHz.[1] Thus the traffic from medical BAN can be off-loaded from the already saturated ISM band of 2400–2500 MHz. Second, by shifting or rotating the offsets of the beacon periods, the coordinator can switch the scheduled slot allocation. Consequently, the impact of interference can be further reduced. Third, the standard provides a dynamic channel hopping method which allows the network to escape from the narrowband interference from other systems. Finally, the standard has also specified a two-hop relaying by using a single relay node. This mechanism can be used when the basic star network topology with single-hop transmissions only cannot provide the necessary levels of reliability. For example, when a person is blocking the LOS path between a pair of transceivers,

---

[1]The 2360–2400 MHz frequency range is available on a secondary basis. As this spectrum is primarily used for aeronautical telemetry, usage of the this frequency band is restricted to indoor operation at health-care facilities and are subject to registration and site approval.

the link may be attenuated too much due to the propagation obstruction to keep the received signal power above the receiver sensitivity [120, 151].

In IEEE 802.15.6, energy utilization is increased via the mechanism of low-power sleep mode. Sensor nodes can stay in the sleep mode for a long time (e.g., for a number of beacon periods) before transmission/reception. Boulis et al. [17] presented four MAC techniques that can be used to increase both the 802.15.6 system reliability and the energy utilization.

In summary, the 802.15.6 standard offers three PHY options and the hybrid mode including different channel access scheduling methods by controlling their lengths and places in a superframe. Therefore, it gives device manufacturers much flexibility in a way to select the working mode which can satisfy the requirements on cost, reliability, energy, etc., and make the tradeoff among these features according to the target application scenario.

(4) The BAN Wireless Channels

Accurately modeling the WBAN channels is vital for researchers to evaluate the network performance in realistic environments. For example, while the WBAN signals are sent from one sensor to a coordinator, the signals propagating through a body will experience attenuation, diffraction, and reflection around the body. The power fading and temporal spread of the signals will degrade the reliability and rate of data transmissions, in particular when sensors are located on various positions on a body. As shown in [147], it is preferable that the packet error rate (PER) is smaller than 1 %.

In the last decade, extensive efforts have been made to characterize the BAN channels based on both measurements and simulations, in order to predict link level performance and develop more effective antennas. These works have been conducted in both the ISM bands around 400 MHz and 2.45 GHz and the UWB spectrum between 3.1 and 10.6 GHz. In all of the bands, intra, on, and off-body propagation environments are investigated [47]. Significant progress has been made, such as the statistical models for the fading of BAN links. The channel models for different body movements and poses with sparse or rich scatterers around have been proposed. The multipath effects on signal transmissions have also been investigated [122]. For example, the measurement results in [19] have demonstrated the path loss for different body positions and frequency bands.

It has been shown that, compared with the UWB signals through the human body conduction systems, narrowband wireless communication is debatably suitable for the medical care applications. Within the spectrum of typical narrowband BANs, the radio channel is essentially flat and slow fading, and the intersymbol interference caused by multipath may be insignificant [6, 17]. The narrowband propagation channel inside a body may be modeled by, for instance, the Rician fading with an appropriate K-factor. The K-factor is the ratio between the average power of the direct propagation path and that of the scattered/reflected paths, and it denotes the channel fading level [113].

In the last decade the UWB channel modeling has drawn great interest and the standardization bodies have put a lot of effort due to the much wider available bandwidth. The standardized channel impulse response (CIR) models for UWB channels have been developed, and typical modulation schemes have been evaluated based on the channel models. For example, the authors in [154] conducted the field measurements on a human body over 3.1–10.6 GHz in an indoor environment and also in an anechoic chamber, and the path loss exponents under different conditions were obtained.

By considering the attenuation, multipath effect, interference, and mobility in wireless body channels, more applicable network structure, MAC protocols, and routing mechanisms in BANs can be developed.

2.4 Summary

The advancements in low-power, large-scale integrated circuits, and wireless communications have promoted the popularity of local, personal, and body area networks, which allow inexpensive and continuous connectivity. This chapter overviews the most important features of the network architecture, PHY, MAC, and resource management defined in the relevant standards.

The 802.11 specifications and a series of amendments have standardized a set of technologies for WLANs. The signal transmissions with the bandwidths of 5, 10, 20, and 40 MHz over multiple channels in the unlicensed 2.4, 5, 45, and 60 GHz and the sub 1 GHz frequency bands have been defined. Based on the PHY schemes, the MAC protocols for channel access and resource scheduling are specified, including the DCF, PCF, EDCA, and HCCA. The LANs based on 802.11 have successfully supported a wide range of applications from data service such as E-mail to multimedia streaming such as IPTV. Driven by customers' manifold demands and advancement of radio technologies, the expansion of the 802.11 products will continue, for example, for very high throughput and dense networks. The very high throughput (VHT) Study Group of IEEE 802.11 is working on the next-generation WLAN standards.

IEEE 802.15, Bluetooth, ZigBee, etc. have been designed for WPANs to cover a range of 10 m. IEEE 802.15.3c operates at the mmWave band and is supposed to provide the data rate over 1 Gbps at the MAC layer. The range of 60 GHz signals is very limited due to high propagation attenuation at this high frequency band. Therefore, repeaters are necessary, and the link-layer mechanisms for multihop relay need to be developed for the deployment of mmWave networks. Furthermore, other technologies such as advanced channel coding and steerable antennas with high directional gain can be employed to improve the coverage. Considering the current advancement of the 60 GHz systems, we can predict that new efficient MAC protocols for resource management are needed to fully exploit the new characteristics of the PHY and will consistently play an important role in the development of next-generation WPANs.

As the last-meter access to the Internet, WBAN is a critical part in providing ubiquitous coverage to users. Located in the close proximity of human bodies,

WBANs have enabled a variety of applications, such as medical care, personal entertainment, automatic workspace, cable replacement, etc. In the last decade, the research on WBANs has been an active topic, including the application scenarios, sensor/actuator devices, radio transceivers, wireless channel models, and interconnection/coexistence of WBANs. While the WBAN technologies have realized the ubiquitous coverage, there are still many open issues to be addressed, particularly for the network interconnection and resource management.

The development of the WLANs, WPANs, and WBANs is driven by the interdisciplinary research interest and great efforts from the academia, industry, and standardization bodies such as the IEEE 802 working group. New challenges need to be met in the development of the wireless networks, including

- Extreme energy efficiency,
- Unique characteristics of the wireless channels,
- Inter-operability and seamless connectivity,
- Managing interference,
- Application requirements, and
- System and device-level security.

While these technologies are currently in the primitive stage, they are undergoing rapid development. A preliminary introduction of the network standards is presented in this chapter, which provides a source for quick understanding of the key ideas, methodologies, and technologies of these networks and their potential applications. The interested readers are referred to the mentioned references for a more comprehensive treatment.

Chapter 3
Contention-Based Medium Access Control

Distributed resource management has several advantages such as simplicity, scalability, and robustness against topology dynamics thanks to its distributed nature, as discussed in Chap. 1. Most distributed MAC schemes are based on the principles of carrier sense multiple access (CSMA). In this chapter, we first introduce in detail the operation of the basic traditional CSMA-based channel access. Then the prioritized CSMA-based medium access scheme is discussed, and the built-in mechanisms for service differentiation such as differentiated arbitration interframe space (AIFS) and CW size are investigated. The prioritized channel access is simulated by OPNET, and the performance in terms of frame delivery delay and per-user throughput is presented. The effectiveness of the differentiation mechanisms is shown by the network simulation results. Due to the randomness in contention, these mechanisms currently provide soft (statistical) QoS guarantee only. In addition, theoretical analysis on the stochastic channel access behavior can give insight into resource management, predict network performance, and accelerate protocol design and optimization process. Three typical analytical approaches for the contention-based MAC are introduced, namely the Markov model, mean value analysis, and backoff counter distribution analysis. All the models study the channel access process of competing stations based on the decoupling-type approximation and fixed-point iteration. The first two methods both calculate the "per-slot" statistics, while the third one describes the steady-state distribution of the backoff counter. These models can accurately predict not only network performance metrics such as throughput and frame delay, but also frame-level performance metrics such as the per-slot transmission and collision probabilities, backoff counter distributions, time interval between two consecutive transmissions, etc.

© The Author(s) 2017
R. Zhang et al., *Resource Management for Multimedia Services in High Data Rate Wireless Networks*, SpringerBriefs in Electrical and Computer Engineering, DOI 10.1007/978-1-4939-6719-3_3

3.1 Channel Access Mechanism

3.1.1 Traditional Contention-Based Medium Access Control

WLAN is a kind of shared-medium communication networks that exchange information over wireless links among all stations. Distributed resource management has
been adopted in WLANs thanks to its simplicity, scalability, and robustness against
topology dynamics. An example of distributed MAC protocols is the DCF of IEEE
802.11 [124] which is the de facto MAC protocol used in WiFi devices. Another
example is the elimination-yield mechanism in HIPERLAN/1.

Distributed MAC protocols target sharing resources fairly among an unknown
number of stations while minimizing access delays and maximizing the overall
throughput. Starting from the pure ALOHA protocol, significant progresses have
been made improving the throughput and stability of CSMA-based MAC protocols.
The IEEE 802.11 DCF protocol adopts the CSMA mechanism by sensing the
carrier in the wireless channel before transmission. Furthermore, the current IEEE
802.11-based stations are half-duplex, as they cannot listen to the channel (sense
the carrier) while transmitting. Hence collisions cannot be detected in the radio
environment as in the case of wired networks (e.g., over twisted copper wires
of Ethernet). Therefore, an 802.11 device waits for a backoff interval before
each frame transmission for collision avoidance (CA). A major advantage of the
CSMA/CA method, which is adopted by DCF, is that the channel access procedure
is simple and easy to implement, because centralized controller and stringent
timing on the radio interface for synchronization are not required. Therefore, the
CSMA/CA mechanism provides easy support for mobility and network scalability,
from classical WLANs to emerging people-centric networks as discussed in Chap. 2.
On the negative side, CSMA/CA can only provide a best-effort transmission service
without QoS guarantee. In this subsection, we follow the specification of DCF to
introduce the general ideas and operations of CSMA/CA.

(1) Three Fundamental Rules of DCF

1. Sense the channel before transmitting. A station needs to check whether the
   medium is idle or not before it tries to send out a frame. For a new frame to be
   transmitted, the station senses the medium and if it is idle for a duration of Distributed (coordination function) InterFrame Space (DIFS) (e.g., 50 μs in IEEE
   802.11b with DSSS PHY), the station can send out the frame. However, if the
   channel is busy or becomes busy during DIFS (another station is transmitting),
   the station needs to backoff by waiting for a random delay as defined in the next
   rule.
2. Use a backoff algorithm to avoid collisions. If the channel is sensed busy or
   the current transmission fails (e.g., due to collision), the station performs a
   random backoff and then senses the channel again. The random waiting time
   is determined by a *backoff counter*.

3. Use a positive acknowledgement scheme to ensure reliability. If the destination has successfully received its frame, the destination station is required to reply an Acknowledgement (ACK) frame to inform the source station. Otherwise, if no positive ACK frame is replied from the destination, the sender becomes aware of the failure of the transaction and will begin backoff for retransmission.

(2) Operations of a Transmitting Station

1. Transmitting a frame:

   - When a station is idle and a new frame arrives (this frame does not follow another one in a burst of frames), it senses the channel activity. If the channel is idle and remains so for a duration equal to DIFS, it transmits the frame and waits for the ACK.
   - When a station is in a backoff stage and the backoff counter reaches 0 (i.e., the backoff procedure finishes), the station transmits the frame and waits for the ACK.

2. Performing backoff:

   - When a station is idle and a new frame arrives, it senses the channel activity for DIFS. If the channel is busy, the station performs backoff by picking a random number as the initial backoff counter value in an exponentially increasing contention window.
   - When a station just completes the transmission of a frame and has a backlogged frame to send, it immediately performs a new backoff procedure no matter whether the medium is idle or not. This policy avoids the same station dominating the channel for a lengthy period.
   - When a frame is sent out but no ACK is received, the station assumes that a collision has occurred. It performs backoff before retransmission.

(3) Operations of a Receiving Station

If a frame is successfully received, the receiver returns an ACK after a short interframe space (SIFS) time. Please note that the SIFS plus the propagation delay is smaller than DIFS so that no other station is able to transmit a new frame before the ACK. In other words, the other stations cannot interrupt the ongoing frame and ACK transmissions.

(4) Rules of Backoff Operation

1. In the first backoff stage of a new frame, a station draws an integer randomly in the range of $[0, W_0 - 1]$, where W_0 stands for the initial CW size. W_0 equals $aCW_{min} + 1$ and aCW_{min} is defined in the 802.11 standards. This drawn integer is the initial value of the *backoff counter*.
2. The station keeps sensing the channel. If the channel has been idle for a DIFS, the backoff process starts. If the current backoff counter equals 0, the station sends out the frame immediately. If the backoff counter is larger than 0, the backoff counter is reduced by 1 if the medium is detected idle for every duration of a

slot time (e.g., *aSlotTime* = 20 μs in 802.11b). Otherwise, the backoff counter should be "frozen" if the channel is sensed busy until the channel is sensed idle for DIFS again (i.e., the on-going over-the-air transmission is finished). When the decrement of the backoff counter resumes, the station continues to decrease the backoff counter by 1 after every idle time slot.

3. The station transmits the entire frame when the backoff counter is reduced to be 0, and waits for an ACK.

4. If a frame transmission fails, presumably due to collision, the sender will not receive an ACK before timeout. The timeout period is predefined. Then it starts the next backoff stage. When a frame transaction fails and the current CW size is smaller than $aCW_{max} + 1$, the CW size is doubled, i.e., $W_{k+1} = 2W_k$, where k is the retransmission (or retry) time which also infers the backoff stage. A new initial value of the backoff counter is selected uniformly in the range of $[0, \ W_{k+1} - 1]$. Then a new backoff stage begins and the backoff counter is decremented as described above. This process will be repeated until the frame is successfully delivered or is discarded when the maximum times of retry are achieved.

5. However, when the CW size W equals $aCW_{max} + 1$, it does not ascend any more and instead remains the value until the retry limit is reached and the frame is discarded. Setting an upper bound for W has the advantage of restricting the maximal frame transmission latency. For a new frame, the backoff process always starts with the first backoff stage, i.e., W is reset to $aCW_{min} + 1$.

As discussed above, in the kth backoff stage, the period of idle channel can be calculated by (the transmissions from other stations result in extra waiting time):

$$Backoff \ Time = Initial \ value \ of \ backoff \ counter \times aSlotTime, \qquad (3.1)$$

where *Initial value of backoff counter* is a random integer drawn uniformly in the interval of $[0, \ W_k - 1]$ and *aSlotTime* is a constant time period defined as a PHY parameter in the 802.11 standard. The initial window size is $W_0 = aCW_{min} + 1$. In IEEE 802.11b, the direct sequence (DS) PHY characteristics of $aCW_{min} = 31$ and $aCW_{max} = 1023$ are specified and the time slot parameter is $aSlotTime = 20 \, \mu s$.

The evolution of the CW size is shown in Fig. 3.1. This exponential backoff mechanism can make one station wait longer on average after each collision, and multiple stations defer and wait for random periods of time after the collision. The station which has the smallest backoff counter will reach zero first and win the contention. Furthermore, the expectation of the waiting time is doubled (or keeps constant if it is large enough) for every retransmission. Thus the probability of another collision among the competing stations is reduced after every collision thanks to the larger CW sizes, which is the collision avoidance feature of the protocol.

As illustrative examples, the operations of the DCF-like contention-based channel access are demonstrated in Fig. 3.2. For easy presentation, the channel access behaviors of two stations are illustrated, which can be described in three scenarios.

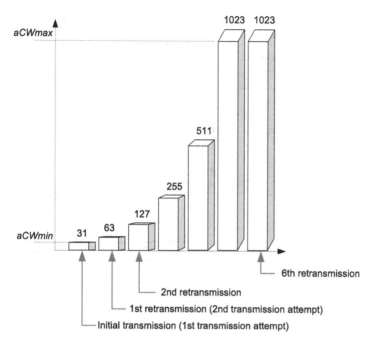

Fig. 3.1 An example of the exponential increase of CW size (direct sequence spread spectrum (DSSS) PHY specification for the 2.4 GHz band designated for ISM applications in IEEE 802.11 [124])

- **Successful Frame Transaction**

  The basic channel access is simple, as shown in Fig. 3.2a. At the beginning, Station A is idle and then a new frame A1 from the upper layer arrives at the MAC sublayer to be transmitted. Station A senses the channel and discovers that the channel is idle for DIFS. Then it sends out Frame A1 immediately. Without collision, the frame is received successfully by its targeted station. The receiver returns an ACK frame after SIFS. Since Station A receives the ACK, it is aware of the successful delivery of Frame A1.

  Suppose that, during the transaction of A1, a new frame, denoted by A2, has arrived at the MAC sublayer from the upper layer. Since Station A is busy in processing Frame A1, Frame A2 is put into its buffer.

  When Frame A1 is delivered, Frame A2 becomes the head-of-queue one and begins to compete for transmission opportunity. Since the transmission of A1 and the arrival of A2 are overlapped, Station A cannot send frames continuously. Therefore, it performs backoff at this moment. As mentioned above, it randomly draws an integer from the $[0, CW_0]$, for example, 5, as the initial value of its backoff counter. Then the backoff procedure starts and Station A keeps sensing the channel. After the channel is sensed idle for DIFS, it decrements the backoff counter by 1 for every idle time slot. After five idle time slots, the backoff counter is reduced to 0 and Station A transmits Frame A2.

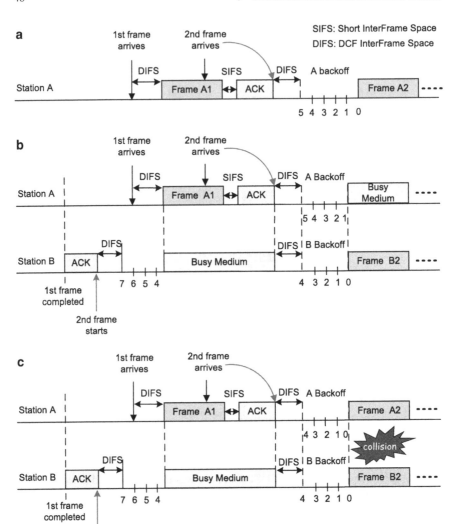

Fig. 3.2 The basic operation of DCF. (**a**) Successful frame transaction. (**b**) Successful collision avoidance. (**c**) Collision occurrence

- **Successful Collision Avoidance**

  Now we consider the scenario that multiple stations are competing for channel access. For easy presentation, there are two stations. As depicted in Fig. 3.2b, Station B has a frame B2 to send. Since Frame B2 is consecutive with B1, Station B performs backoff with an initial backoff counter value of 7. Its backoff counter is reduced by 1 after every idle time slot, as mentioned above. However, Frame A1 is sent out from Station A and thus the channel is sensed busy in the fourth time slot. As a result, the backoff counter of Station B is frozen during the transaction of Frame A1 including the transmissions of both the data and ACK frames.

Please note that Station B will not decrement the backoff counter even when the channel is idle during the SIFS between the data frame and the ACK frame. This is because Station B must sense the channel idle for DIFS before assessing that the channel is idle. As defined in the DCF specification, SIFS is smaller than DIFS. Thus the channel is busy again (due to the transmission of the ACK frame) before the channel keeps idle for DIFS. Station B will not start/resume backoff or transmit a frame during SIFS and hence the transaction of Frame A1 will not be interrupted. After the transmission of the ACK frame, the channel is sensed idle for DIFS, and Station B resumes decrementing its backoff counter. Starting from the value of 4, the backoff counter is reduced to 0 after four idle time slots. Then Station B sends out Frame B2.

Similarly, when Station B is transmitting Frame B2, the channel is sensed busy by Station A so the backoff counter of Station A is frozen to avoid collision with Frame B2. Again, after the transaction of Frame B2, the channel can be sensed idle for DIFS, and Station A resumes decrementing its backoff counter. Thus, collisions between the two competing stations are avoided.

- **Collision Occurrence**

Although collisions may be mitigated by using carrier sensing and backoff, collisions can still happen in the contention-based MAC. As illustrated in Fig. 3.2c, if the initial value of the backoff counter of Station A for Frame A2 is 4, then the two stations reduce their backoff counters to 0 simultaneously. Consequently, they will transmit frames in the same time slot and a collision will occur.

Since the frames cannot be received correctly, the targeted receivers will not reply ACK frames. After timeout, Stations A and B will perform the next backoff stage. They will double their CW sizes as described above (exponential backoff) and draw the initial values for their backoff counters randomly. If their initial values are different, the transmission time of the two stations will be different. Thus their frames will not collide.

Otherwise, if they coincidentally select the same value from the CWs, they will retransmit simultaneously and collide again. Then the two stations will perform the next backoff stage. Since every time when a collision occurs, the CW sizes are doubled. The probability that the two stations select the same integer will be decreased exponentially. Therefore, the collision probability will be reduced after every collision, which is an effective mechanism to resolve the collisions among stations.

In a WLAN with heavy traffic load and a large number of stations, a frame may encounter collisions even when the CW size is quite large. In this case, the frame will be discarded when it has been retransmitted for a maximal number of times.

As we can see, since the MAC parameters such as *DIFS*, aCW_{min}, and aCW_{max} are fixed for all traffic flows of all stations, the flows are not differentiated and have the same chance to access the channel. In other words, the priority in accessing the wireless medium is the same for all flows. Therefore, by using the traditional PHY and MAC sublayer mechanisms with fixed parameters for all traffic, the delay, bandwidth, and throughput for all flows are the same statistically.

As discussed above, there is no built-in mechanism to support service priority, guaranteed delay or throughput in the traditional CSMA mechanism. Furthermore, 802.11 DCF expends a considerable amount of airtime in managing channel access such as backoff, collisions, and retransmissions. The collision, retransmission, and stochastic backoff procedure lead to a high variance of the throughput and delay, which is not suitable for delay-sensitive multimedia traffic.

3.1.2   Prioritized Contention-Based Medium Access Control

To better support multimedia traffic, the basic CSMA mechanism can be extended to provide service differentiation. The prioritized CSMA mechanism allows traffic classes with higher priorities (such as the audio/video packets) to access channel earlier. To support QoS in channel resource scheduling, traffic classification and differentiated independent backoff entities need to be defined, as discussed in this subsection.

(1) Access Categories (ACs)

Traffic flows are classified and prioritized according to their QoS requirements. For example, four ACs with eight priorities are defined in IEEE 802.11e (see Sect. 2.1 for details). The ACs are labeled according to their target applications, including *AC_VO* (voice), *AC_VI* (video), *AC_BE* (best effort), and *AC_BK* (background). Usually the delay-sensitive traffic such as real-time voice and video data is assigned high priorities, while the delay-tolerant traffic such as Email and file transfer is assigned low priorities. Different parameter sets including interframe spacing, CW size, and duration of transmission opportunities (TXOPs) are defined and associated with the ACs to provide priorities in medium access contention.

In order to support the differentiated channel access for the four ACs, multiple independent backoff entities should be implemented in one station. MAC service data units (MSDUs) are backlogged in parallel backoff entities which contend for channel access opportunities independently, using the differentiated AC-specific parameters. The architecture for the four parallel backoff entities in an 802.11e station is illustrated in Fig. 2.2 in Sect. 2.1.

(2) Service Differentiation Mechanisms

To provide service differentiation, Ada and Castelluccia [1] proposed to scale the CW size and use a different IFS or maximum frame length for services. Accordingly, the IEEE 802.11 working group has proposed the 802.11e standard to enhance the original 802.11 MAC sublayer to support QoS [54]. The defined EDCA protocol can allow prioritized medium access for applications with QoS requirements. These mechanisms map QoS metrics into the existing basic contention-based MAC parameters, thus avoiding a redesign of the MAC protocol.

Each AC within a single station, as discussed above, should be regarded and behave as a virtual station. The aggregated flow of an AC independently contends for channel TXOP and performs its backoff and transmissions. Thus, prioritized channel access is performed by both internal (intra-station) and external (inter-station) differentiated contention among ACs.

1. *Intra-station contention*: A station may have multiple traffic flows belonging to different ACs. Each priority class has its own queue and backoff counter, as illustrated in Fig. 2.2. Collisions can occur between the flows in different ACs within a single station. When the backoff counters of two or more queues reach zero in the same time slot, these queues have internal collision with each other. In this case, the internal collision is resolved within the station. The frame with the highest priority is granted the chance to be sent. Meanwhile, the other lower-priority frames will perform the exponential backoff and behave as if they had been sent out and had an external collision with other stations. The backoff process is the same as discussed in the previous subsection.
2. *Inter-station contention*: The traffic flows of different stations may transmit frames in the same time slot. In this case, the signals from the stations collide in the common channel. Without ACKs for these frames, the senders will perform exponential backoff.

In both the intra and inter-station contention, the channel access priorities of the competing traffic flows can be differentiated. Higher priority in media access is realized by allocating a smaller waiting time (e.g., smaller IFS) and/or a smaller CW size. Using such strategies, a higher-priority traffic flow can occupy the channel more quickly and thus block the transmission from lower-priority traffic flows. There are three approaches for prioritized contention, as described below.

- **Using Different IFSs for Priority**
  Instead of using a constant DIFS in DCF, different IFSs can be used for different ACs. Using differentiated IFS has been investigated by many researchers because it already exists in the basic contention-based MAC. For example, Deng et al. [33] suggested to use the PIFS and DIFS values in DCF to differentiate two classes of priority. Aad et al. [1] also specified multiple IFSs to differentiate traffic classes. In this mechanism, after a busy channel slot with any transmission activity, a higher-priority flow waits for a shorter IFS to start or resume counting down its backoff counter, while a low priority flow waits for a longer IFS,

deferring the countdown for an additional time period. Hence a high-priority flow decrements its backoff counter earlier and can obtain the transmission opportunity earlier on average.

The EDCA proposals specified in IEEE 802.11e have introduced a new type of IFS named arbitration interframe space (AIFS) [54]. Instead of using a constant DIFS in DCF, different AIFSs are adopted. In EDCA, various $AIFS[AC]$ values are defined for ACs and differ from each other for an integer number of backoff slots. Following DCF, if the channel is sensed to be idle for $AIFS[AC]$, the flows belonging to the corresponding AC can start/resume their backoff stages.

The smallest AIFS is DIFS, and it is used for the highest priority traffic. AIFS is enlarged gradually for the other ACs according to the priorities. The higher the priority is, the smaller the AIFS is. Because the waiting time for an idle channel, $AIFS[AC]$, varies among ACs, a flow belonging to a high-priority AC needs to wait for a shorter AIFS. Consequently, it will start backoff/transmission earlier, which leads to a better chance to obtain the TXOP. Real-time multimedia traffic flows (such as video and voice) are assigned shorter AIFSs, so they obtain higher priorities in accessing channel and smaller delay.

The AIFS value is computed by

$$AIFS[AC] = AIFSN[AC] \cdot aSlotTime + aSIFSTime, \qquad (3.2)$$

where $AIFSN[AC]$ is an integer and $aSlotTime$ is the duration of a backoff time slot. The default values of $AIFSN[AC]$ defined in the EDCA Parameter Set are listed in Table 2.3 in Sect. 2.1. It is specified that $AIFSN[AC] \geq 2$ for normal stations and $AIFSN[AC] \geq 1$ for APs. Consequently, $AIFS[AC]$ for contending stations cannot be smaller than the $DIFS$ defined in DCF which equals to $2 * aSlotTime + aSIFSTime$. In a network using EDCA, the AP can adjust the values of $AIFSN[AC]$ according to the network conditions, and then broadcasts the results chosen for each AC in each beacon frame. However, such parameter setting leads to a potential problem. Because the new AIFS values are longer than the existing DIFS, the frame of a station using the current DCF scheme will automatically acquire the highest priority when competing with stations using the EDCA mechanism.

In EDCA, the backoff counter is resumed one time slot before the expiration of its class AIFS. Thus, the backoff counter equals the previous frozen value reduced by one at the end of the AIFS. Then a transmission is scheduled in the time slot following the one where the backoff counter is decremented or resumed to zero.

Figure 3.3 gives an illustration of the channel access process and timing with differentiated $AIFS[AC]$. Three priorities are shown in the figure.

- **Using Different Contention Windows for Priority**
  To further differentiate the channel access chance among ACs, CW sizes also vary for different ACs. As described previously, the backoff counter corresponds to the number of idle time slots that a station needs to wait for after the IFS. The idea of CW differentiation is that, an AC with a higher priority uses smaller CW sizes in the random selection of the initial value of the backoff counter.

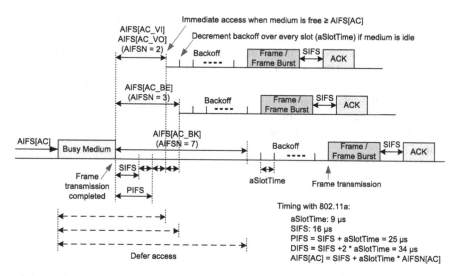

Fig. 3.3 The operation of prioritized contention with differentiated AIFS

For example, for two traffic classes denoted by A and B, there are two ranges of CW sizes: W_A (between $W_{A,min}$ and $W_{A,max}$) and W_B (between $W_{B,min}$ and $W_{B,max}$). As proposed by Chesson et al. [25] and Benveniste [10], the values of CW size are assigned such that the W_{min} and W_{max} of a low-priority class are larger than those of a high-priority class. Since the backoff counter is a random number uniformly distributed between W_{min} and W_{max}, the lower-priority traffic will select a larger value on average, and vice versa. Thus, a flow with a higher priority needs to wait for fewer idle time slots on average. Consequently it has a larger probability to finish backoff process earlier and sends frames before lower-priority flows. Therefore, the higher-priority frames have a better chance to obtain the TXOPs. The ranges of the CW sizes can either overlap or not. If the CW ranges are completely separated without overlapping, the higher-priority traffic has higher chances to have transmissions before the lower-priority traffic. Varying CW sizes for different ACs are realized by setting $aCW_{min}[AC]$ and $aCW_{max}[AC]$ for the CW size limits. For example, the default values of the minimal and maximal CW sizes defined in the EDCA Parameter Sets are listed in Table 2.3 in Sect. 2.1. Based on the $aCW_{min}[AC]$, the window size for each backoff stage is computed.

However, this solution may encounter the starvation problem. As the high-priority traffic load increases, it tends to grab the channel persistently, preventing the access by low-priority traffic.

(3) Distributed Fair Scheduling (DFS)

Although binding the bandwidth to high-priority traffic is desirable for QoS provisioning, it also results in unfair resource allocation and starvation of low-priority traffic. Fair queuing mechanisms have been proposed to solve this issue, aiming to ensure that the throughputs of flows in two traffic classes are in a given ratio. With the mechanism used in [150], each traffic class gets the appropriate portion of the bandwidth by regulating the waiting time, which is different from the schemes that bind the channel access to priority.

Distributed fair scheduling (DFS) proposed by Vaidya et al. [130] maps traffic classes into the backoff intervals instead of fixing the CW size ranges for different priorities. The main idea of this scheme is to randomly select a backoff interval which is positively relevant to the ratio of the packet length and the weight of a frame.

There are also other proposals for incorporating QoS mechanisms with distributed protocols. In [131], the authors proposed the virtual MAC and virtual source. A node could differentiate the resource allocation for voice, video, and data. In [119], the authors proposed a scheme that split the transmission period into two parts, one for real-time traffic and the other for non-real-time traffic. Thus, the transmission opportunities for the former are guaranteed for QoS support, while the basic contention-based MAC scheme was dramatically changed.

We focus on the EDCA-like prioritized MAC in the rest of this chapter because it has been standardized by IEEE. The IEEE 802.11e EDCA [54] is a combination of the approaches discussed above, using different IFSs and CW sizes to provide service differentiation and priority.

3.2   Performance Evaluation

3.2.1   Simulation Settings

In this section, the performance of the differentiated contention-based MAC mechanism is evaluated with OPNET simulation. In the simulation, all IEEE 802.11 wireless stations with EDCA parameters are configured in the ad hoc mode. All the stations are located on a circle with the radius of 75 m, as shown in Fig. 3.4 where the number of stations is changed from 6, 12, 18, to 24. Two stations form a pair and they transmit frames to each other. Thus each station is a sender and a receiver simultaneously. The two stations in a pair are placed at the ends of a diameter. When there are N stations in the network, they are grouped into $\frac{N}{2}$ pairs. Since the network is in the single-hop, ad hoc mode, the frames are transmitted directly from the source to destination station. All stations remain stationary during the simulation. Due to the symmetric topology, the physical environments and parameters, such as the signal strength and capture effect, of all the stations are identical. Thus the physical-layer characteristics can be ignored in comparing the performance of different flows.

In the simulation, we define that there are three ACs (channel access levels), named by AC(0), AC(1), and AC(2) in the order of increasing channel access priorities. Two stations in a pair belong to the same AC. Thus, a total of $\frac{N}{3}$ stations ($\frac{N}{6}$ pairs) are in one AC using the corresponding contention parameter set.

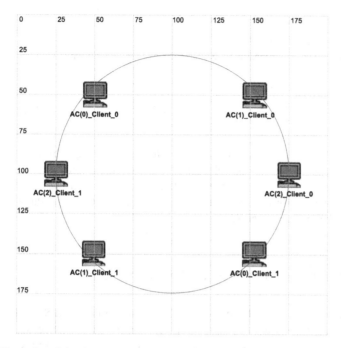

Fig. 3.4 Illustration of the simulation network topology. The number of stations can be changed to be 6, 12, 18, and 24

A standard OPNET 802.11b PHY module using DSSS scheme with maximum data rate up to 5.5 Mbps is adopted. The PHY characteristics are set as follows: $aSIFSTime = 10\,\mu s$, $aSlotTime = 20\,\mu s$, $aLongRetryLimit = 7$, and the TXOP limit of 3.264 ms. We simulate the EDCA access process with varying AIFS and CW size to evaluate their impact on the resource management and QoS provisioning.

Traffic (frames) of the three ACs are fed into the MAC layer from the higher layer. The frame payload size of each AC is random in the range of [1000, 2000] bytes with uniform distribution. In order to evaluate the effect of contention parameter differentiation, the stations are saturated, i.e., there are always frames backlogged in their buffers for transmission. Each station has a buffer of 32, 000 bytes for frame queueing.

The performance metrics evaluated are:

1. *End-to-end delay.* This delay includes the queuing delay in the source buffer and the medium access time (i.e., the frame service time) for transmission over the wireless channel. The queueing delay is the time interval from the moment a frame arrives at the buffer (MAC service point) to when it becomes the head of the queue. The service time is the interval when the frame begins to contend for channel access to when it is successfully received. The delay for data frames successfully received by the MAC and forwarded to the higher layer are also

Table 3.1 The AIFSN parameter set for the ACs

Access category	AIFSN				
(AC)	Scenario 1	Scenario 2	Scenario 3	Scenario 4	Scenario 5
0	7	7	7	7	7
1	7	6	5	4	3
2	6	5	4	3	2

counted. The average end-to-end delay refers to the average result of the end-to-end delay of all the stations belonging to one AC in the simulation.

2. *Station throughput.* This is the per station throughput in the simulation. The average station throughput is the average result of the throughput of all the stations belonging to one AC.

3.2.2 Prioritized Access by Differentiated AIFS

In this subsection, we evaluate the impact of differentiated AIFS on the channel access behavior and resource allocation among the ACs. The traffic label and channel access parameters except AIFSN of all the three ACs are the same, and the values of the AIFSN parameter are configured differently for five scenarios, as listed in Table 3.1. The AIFSN of AC(2) is the smallest in each scenario, and it should have the highest priority in channel access. The AIFSN of AC(0) is always 7 that is the largest, and thus AC(0) has the lowest priority. AC(1) has the medium priority. Meanwhile, the AIFSNs of AC(1) and AC(2) decrease from Scenario 1 to 5, and consequently the difference between the AIFSN values of AC(0) and AC(2) becomes larger. All the ACs adopt the same CW size in the exponential backoff, where the minimal and maximal CW sizes are $aCW_{min} = 15$ and $aCW_{max} = 1023$, respectively.

Figure 3.5 plots the instant end-to-end delay and the throughput of a station of AC(0) in different scenarios. There are a total of 18 stations (as mentioned earlier, six stations/flows belong to each AC). The horizontal axis represents the simulation time in seconds. From Scenario 1 to 5, as the AIFSNs of AC(1) and AC(2) decrease, the channel access behavior/statistics are further differentiated. Since AC(0) has the lowest priority, the allocated channel resource and its TXOPs, are further reduced. As the result, we can see the tendency that the throughput of the AC(0) station decreases while its service delay becomes longer. For presentation clarity, the average of the end-to-end delay and throughput of all the stations belonging to AC(0) are plotted in Fig. 3.6.

To further demonstrate the effectiveness of the differentiated AIFS values, the performance of the three ACs is compared. The average end-to-end delay and throughput of the ACs are plotted in Fig. 3.7. As an illustrative example, the

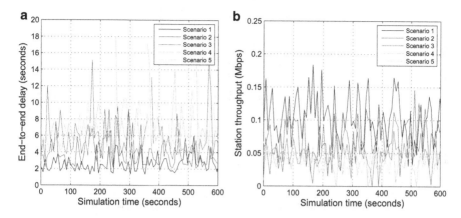

Fig. 3.5 The evolution of the delay and throughput of a station of AC(0) for the five scenarios of AIFSN configurations (total number of stations: 18). (**a**) Frame end-to-end delay. (**b**) Station throughput

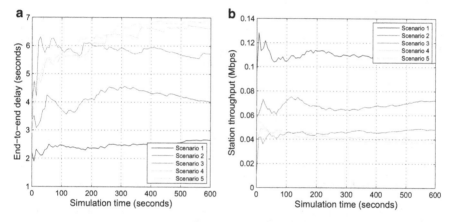

Fig. 3.6 The evolution of the average delay and throughput of all stations of AC(0) for the five scenarios of AIFSN configurations (total number of stations: 18). (**a**) End-to-end frame delay. (**b**) Station throughput

parameter setting in Scenario 3 is adopted and there are a total of 18 stations. As can be observed, the performance of the three ACs is effectively differentiated and the channel access priorities are provided.

In order to show the performance as a function of the number of stations, Fig. 3.8 plots the average end-to-end delay and throughput for the three ACs with the parameter set of Scenario 3 given in Table 3.1, and the number of stations (flows in the network) increases from 6 to 24. For example, when there are six stations in the network, actually one pair (two stations) with two flows exists in each AC. Similarly, the network of 24 stations includes four pairs and eight flows in each AC.

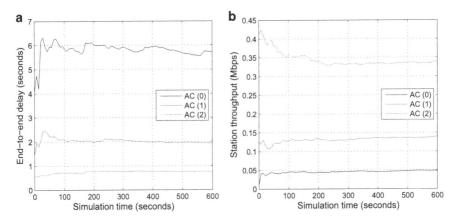

Fig. 3.7 The evolution of the average delay and throughput of the three ACs (total number of stations: 18; scenario index: 3). (**a**) End-to-end frame delay. (**b**) Station throughput

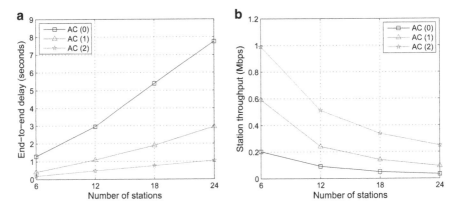

Fig. 3.8 The performance of the three ACs with respect to the number of stations (scenario index: 3). (**a**) End-to-end frame delay. (**b**) Station throughput

As can be seen in Fig. 3.8, AC(2) has the smallest average end-to-end delay, and AC(0) has the largest. Meanwhile the collision rate in channel access rises resulting from the increased number of stations. It is clear that AC(2)'s delay increases slightly, while the AC(0)'s and AC(1)'s delay increases more significantly. Consequently, the throughput of AC(2) decreases relatively less than the other ACs. The results show that varying AIFSN is effective in differentiating the channel access priority. In addition, when the network load is increased, the highest-priority traffic class can be protected and its channel access performance is not degraded much, while the lower-priority traffic classes are more affected.

Figure 3.9 plots the average end-to-end delay and throughput of the three ACs for the five scenarios defined in Table 3.1, and the number of stations equals 18. As expected, AC(1) and AC(2) obtain more chance than AC(0) to access the channel

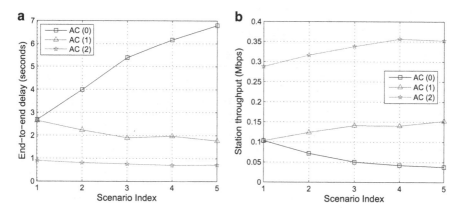

Fig. 3.9 The performance of the three ACs with different AIFS configurations (total number of stations: 18). (**a**) End-to-end frame delay. (**b**) Station throughput

Table 3.2   The CW size parameter set for the ACs

Access category	$[aCW_{min}, aCW_{max}]$			
(AC)	Scenario 1	Scenario 2	Scenario 3	Scenario 4
0	[31, 1023]	[63, 1023]	[127, 1023]	[255, 1023]
1	[15, 1023]	[31, 1023]	[63, 1023]	[127, 1023]
2	[15, 1023]	[15, 1023]	[15, 1023]	[15, 1023]

resource. The average frame service time of both AC(1) and AC(2) decreases because their AIFSNs are reduced. The performance of AC(0) deteriorates although its parameters remain the same. This is because the AIFSNs of AC(1) and AC(2) are both reduced and they obtain even more chances to win the contention and access the channel. Since the channel resources are more occupied by these two ACs, AC(0) has fewer transmission opportunities and thus its performance deteriorates considerably.

3.2.3   Prioritized Access by Differentiated CW Size

In this subsection, we evaluate the impact of differentiated CW size on the channel access priority. The values of the channel contention parameters in this simulation set are listed in Table 3.2. The stations are still divided into three ACs, and the traffic label and channel access parameters except the CW size of all the three ACs are identical. Furthermore, the CW sizes vary in four scenarios. The initial CW size (in the first backoff stage) is aCW_{min}, and it is doubled after every transmission failure until the maximal CW size, aCW_{max}. The maximal CW size is fixed as $aCW_{max} = 1023$ for all the three ACs in order to avoid too long delay for a frame. But the

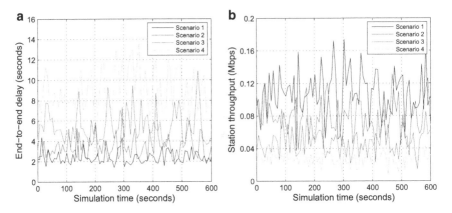

Fig. 3.10 The evolution of the delay and throughput of a station of AC(0) for the four scenarios of CW size configurations (total number of stations: 18). (**a**) Frame end-to-end delay. (**b**) Station throughput

minimal CW size aCW_{min} is varying. The CW size of AC(2) is the smallest in each scenario, which is in the range of $[aCW_{min}, aCW_{max}] = [15, 1023]$. Therefore, it should have the highest priority in channel access. The aCW_{min} of AC(0) is the largest in the four scenarios, which is set to be 31, 63, 127, and 255. Hence AC(0) has the lowest priority. The value of aCW_{min} for AC(1) is between the other ACs. The AIFSN of all the ACs equals 7.

Figure 3.10 plots the instantaneous end-to-end delay and throughput of a station of AC(0) in the four scenarios, when there are 18 stations. The horizontal axis represents the simulation time in seconds. From Scenarios 1 to 4, as the CW sizes of AC(0) and AC(1) are enlarged, the stations of AC(2) get more transmission opportunities and thus less channel resources are allocated to AC(0). Consequently, the frame service delays of the AC(0) stations become longer and their throughputs decrease. The average results of the end-to-end delay and throughput of all the stations of AC(0) in the four scenarios are plotted in Fig. 3.11. The channel access performance is affected significantly by the setting of the CW sizes of the ACs.

The effectiveness of the differentiated CW sizes is illustrated by comparing the frame performance of the three ACs. The evolutions of the average end-to-end delay and throughput of all the stations in each AC in the simulation are plotted in Fig. 3.12. There are 18 stations operating in Scenario 3, as listed in Table 3.2. The performance of AC(2) is much better than those of AC(1) and AC(0), due to the differentiated CW sizes.

Figure 3.13 plots the average end-to-end delay and the average throughput for the three ACs with the CW size configuration of Scenario 3 given in Table 3.2, and the number of stations increases from 6 to 24. As can be seen, the end-to-end delay of AC(2) is the smallest and that of AC(0) is the largest, as expected. Because more stations are competing for the channel resource and collision rate increases, the transmission delay increases and the throughput decreases for all the three ACs.

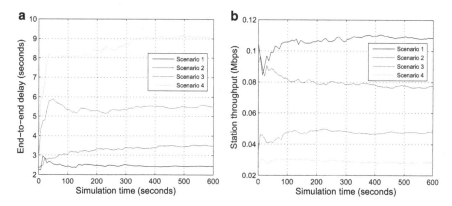

Fig. 3.11 The evolution of the average delay and throughput of all stations of AC(0) for the four scenarios of CW size configurations (total number of stations: 18). (**a**) End-to-end frame delay. (**b**) Station throughput

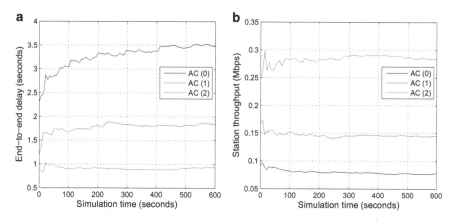

Fig. 3.12 The evolution of the average delay and throughput of the three ACs (total number of stations: 18; scenario index: 3). (**a**) End-to-end frame delay. (**b**) Station throughput

Finally, Fig. 3.14 shows the average performance of the ACs in the four scenarios and there are a total of 18 stations. It is also clear to see that the tendencies of both the end-to-end delay and throughput of AC(2) are opposite to those of AC(1) and AC(0). For example, in Fig. 3.14a, the end-to-end delay of AC(0) and AC(1) increases from the Scenario 1 to 4, but that of AC(2) reduces. This indicates that with the increase of the CW sizes of AC(0) and AC(1), AC(2) not only has the highest priority in channel access but also obtains more channel resources (i.e., TXOPs). Meanwhile, the performance of AC(0) and AC(1) deteriorates. The performance of throughput can be analyzed similarly. Therefore, differentiating the CW size is also effective in providing prioritized channel access.

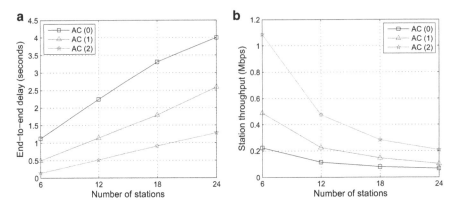

Fig. 3.13 The performance of the three ACs with respect to the number of stations (scenario index: 3). (**a**) End-to-end frame delay. (**b**) Station throughput

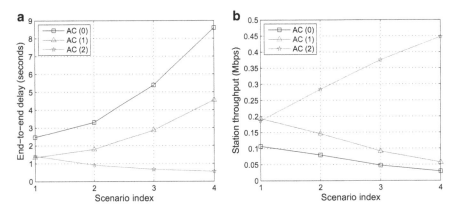

Fig. 3.14 The performance of the three ACs with different CW size configurations (total number of stations: 18). (**a**) End-to-end frame delay. (**b**) Station throughput

3.3 Analytical Modeling for Contention-Based MAC

3.3.1 Analytical Modeling Principles

Rapid deployment of WLANs and the increasing QoS requirements have motivated extensive performance studies of the contention-based MAC protocol in the literature. Analytical models are important for performance prediction, protocol optimization, and deployment of wireless networks. Throughput, delay, and capacity analyses have been studied extensively. Among them, three major streams for analytical models have been proposed, which are the Markov chain-based approach, mean value analysis, and backoff counter distribution analysis. The guiding principle behind these different models is actually common. The main differences are the approaches to finding the collision and transmission probabilities and various choices of renewal cycles, as discussed below. Note that these methods are essentially equivalent and can be transformed between each other.

In this subsection, we discuss the common features of these models, including the basic principles, assumptions, and theories in all the analytical models. Then, the three analytical approaches are introduced in details in the following subsections, including their frameworks and related works. Interested readers are referred to the references mentioned in this section to study more analytical models that have been derived from these basic models.

(1) Homogeneous Access Probability Assumption

From the point of view of system characteristics, using the traditional MAC protocols such as DCF, the flows are *homogeneous* because their backoff parameters are identical. Nevertheless, the prioritized MAC protocols such as EDCA employ various backoff parameters for different ACs. The differentiated CW sizes, backoff multipliers, time to wait before restarting/resuming backoff (i.e., AIFS), or duration of TXOPs result in the heterogeneity among the flows. Therefore, the flows in the whole network are *heterogeneous*. However, the flows belonging to the same AC are still homogeneous as they adopt the same set of contention parameters.

All of the three approaches assume that the stations of the same AC access an idle channel slot with the same transmission probability and collision probability, and they are actually equivalent and interchangeable with each other. The probabilities are assumed constant in the steady state of a network.

For channel access analysis, the probability for the nth station belonging to the lth AC to transmit a frame in a generic slot is a constant (denoted by τ_n). Similarly, the probability for a frame of the nth station to have collision is usually assumed to be a constant too (denoted by p_n). These are key parameters because the other channel access statistics are derived based on them, as discussed in the following.

(2) Renewal Cycle Approach

In different analytical models, the renewal cycle is adopted, as shown in Fig. 3.15. The statistics of the channel access behavior of a station (i.e., backoff counter evolution process) are stationary in different cycles. This results in the fact that the cycles are independent and the channel activity statistics are regenerated in every cycle. The analytical models usually take explicitly or implicitly the time between two adjacent successful transmissions from one station (or flow) as a renewal cycle. The analysis is effective based on the fact that the channel access process for frames repeats among the renewal cycles. It is further assumed that the behavior of a station is homogeneous in each slot inside a renewal cycle. Based on the approximations above, a WLAN using the contention-based MAC is a fixed-point system that can be solved through numerical techniques. The throughput and other performance metrics are derived based on the statistics of the renewal cycles.

At first, suppose that the retry limit of the nth station is $K_n + 1$ (a maximum number of K_n retransmissions can be performed for a frame) and the indexes of the backoff stage are $k = 0, 1, 2, \cdots, K_n$. For example, the short retry limit is 7

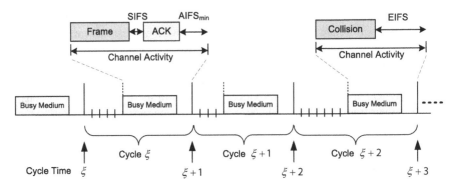

Fig. 3.15 An illustration of the channel access cycles

while the long retry limit is 4 in the IEEE 802.11 specifications. Given that the collision probability of the *n*th station is p_n, the number of transmission trials in a renewal cycle, denoted by R_n, follows a truncated geometric distribution. The average number of the transmission trails, $E[R_n]$, is given by

$$E[R_n] = p_n^0(1 - p_n) + 2p_n(1 - p_n) + \cdots + K_n p_n^{K_n-1}(1 - p_n) + (K_n + 1)p_n^{K_n}$$

$$= 1 - p_n + 2p_n - 2p_n^2 + 3p_n^2 - 3p_n^3 + \cdots + K_n p_n^{K_n-1} - K_n p_n^{K_n} + (K_n + 1)p_n^{K_n}$$

$$= 1 + p_n + p_n^2 + \cdots + p_n^{K_n-1} + p_n^{K_n}$$

$$= \sum_{k=0}^{K_n} p_n^k. \tag{3.3}$$

Let $E[B_{n,k}]$ denote the expected initial value of the backoff counter of the *n*th station in the *k*th backoff stage, i.e., the average number of slots for the station to experience in the stage. Then, the average number of backoff slots of a frame during the whole backoff process, denoted by $E[B_n]$, is calculated as

$$E[B_n] = p_n^0(1 - p_n)E[B_{n,0}] + p_n^1(1 - p_n)\sum_{k=0}^{1} E[B_{n,k}] + \cdots$$

$$+ p_n^{K_n-1}(1 - p_n)\sum_{k=0}^{K_n-1} E[B_{n,k}] + p_n^{K_n}\sum_{k=0}^{K_n} E[B_{n,k}]$$

$$= (1 - p_n)\left[E[B_{n,0}]\sum_{k=0}^{K_n-1} p_n^k + E[B_{n,1}]\sum_{k=1}^{K_n-1} p_n^k + \cdots + E[B_{n,K_n-1}]\sum_{k=K_n-1}^{K_n-1} p_n^k \right]$$

$$+ p_n^{K_n} \sum_{k=0}^{K_n} E[B_{n,k}]$$

$$= (1 - p_n) \left[E[B_{n,0}] \frac{p_n^0 (1 - p_n^{K_n})}{1 - p_n} + E[B_{n,1}] \frac{p_n^1 (1 - p_n^{K_n - 1})}{1 - p_n} + \cdots \right.$$

$$\left. + E[B_{n,K_n-1}] \frac{p_n^{K_n-1}(1 - p_n)}{1 - p_n} \right] + p_n^{K_n} \sum_{k=0}^{K_n} E[B_{n,k}]$$

$$= \left[E[B_{n,0}] p_n^0 - E[B_{n,0}] p_n^{K_n} + E[B_{n,1}] p_n - E[B_{n,1}] p_n^{K_n} + \cdots \right.$$

$$\left. + E[B_{n,K_n-1}] p_n^{K_n-1} - E[B_{n,K_n-1}] p_n^{K_n} \right] + p_n^{K_n} \sum_{k=0}^{K_n} E[B_{n,k}]$$

$$= \left[\sum_{k=0}^{K_n-1} \left(p_n^k E[B_{n,k}] \right) - p_n^{K_n} \sum_{k=0}^{K_n-1} E[B_{n,k}] \right] + p_n^{K_n} \sum_{k=0}^{K_n-1} E[B_{n,k}] + p_n^{K_n} E[B_{n,K_n}]$$

$$= \sum_{k=0}^{K_n} \left(p_n^k E[B_{n,k}] \right) . \tag{3.4}$$

The result in (3.4) is general in the sense that the CW sizes and the distribution of the initial values of the backoff counter in the K_n backoff stages can be arbitrary.

The CW of the nth station in the kth backoff stage is denoted by $[0, W_{n,k}]$ with the size of $W_{n,k} + 1$. As specified in the 802.11 standards, the CW size is exponentially expanded by the persistence factor (PF) f_n in the first $K_n' + 1$ backoff stages (the backoff stage indexes are $0, 1, \cdots, K_n'$) and keeps constant in the remaining $K_n - K_n'$ stages (the backoff indexes are $K_n' + 1, K_n' + 2, \cdots, K_n$). Hence the CW sizes are determined by

$$W_{n,k} = \begin{cases} (f_n)^k W_{n,0}, & 0 \leq k \leq K_n', \\ (f_n)^{K_n'} W_{n,0}, & K_n' + 1 \leq k \leq K_n, \end{cases} \tag{3.5}$$

where $W_{n,0} = aCW_{\min}[AC]$ is the initial CW size for the AC of the nth station (e.g., as listed in Table 2.3 in Sect. 2.1). The 802.11 specifications have defined $f_n = 2$.

Furthermore, it is specified in the 802.11 standards that the initial value of the backoff counter is uniformly drawn from the CW (in the range of $[0, W_{n,k}]$) for the kth backoff stage. Thus the initial value (average number of slots for the station to experience) in the kth stage is

$$E[B_{n,k}] = \frac{1 + 2 + \cdots + (W_{n,k} - 1) + W_{n,k}}{W_{n,k} + 1} = \frac{(W_{n,k} + 1) \frac{W_{n,k}}{2}}{W_{n,k} + 1} = \frac{W_{n,k}}{2}. \tag{3.6}$$

For the 802.11 specification, by plugging (3.5) and (3.6) into (3.4), the average total number of backoff slots for a frame to experience can be further derived as

$$
\begin{aligned}
E\left[B_n\right] &= \sum_{k=0}^{K'_n-1}\left(p_n^k E\left[B_{n,k}\right]\right) + \sum_{k=K'_n}^{K_n}\left(p_n^k E\left[B_{n,K'_n}\right]\right) \\
&= \sum_{k=0}^{K'_n-1}\left(p_n^k \frac{W_{n,k}}{2}\right) + \frac{W_{n,K'_n}}{2}\sum_{k=K'_n}^{K_n} p_n^k \\
&= \frac{1}{2}\left[\sum_{k=0}^{K'_n-1}\left(p_n^k f_n^k W_{n,0}\right) + f_n^{K'_n} W_{n,0}\sum_{k=K'_n}^{K_n} p_n^k\right] \\
&= \frac{W_{n,0}}{2}\left[\sum_{k=0}^{K'_n-1}\left(p_n f_n\right)^k + f_n^{K'_n}\sum_{k=K'_n}^{K_n} p_n^k\right] \\
&= \frac{W_{n,0}}{2}\left[\frac{1-\left(p_n f_n\right)^{K'_n}}{1-p_n f_n} + \frac{\left(p_n f_n\right)^{K'_n}\left(1-p_n^{K_n-K'_n+1}\right)}{1-p_n}\right].
\end{aligned}
\tag{3.7}
$$

(3) Slotted Model for the Backoff Process

The network is modeled as a slotted and synchronized system. On one hand, when the channel is idle, the backoff counters of stations are reduced by 1 after every idle time slot. On the other hand, when the channel is busy, the transaction of a frame (including both the transmissions of the data frame and its ACK frame) is regarded as a time slot. Therefore, no matter when the medium is busy or idle, the channel time is discretized into slots and all stations are synchronized to operate in time slots. Since the durations of the time slots are variable, they are referred to as *generic time slots*.

Under the saturation assumption, all stations are persistently competing for the channel. Thus, no stations may stop contention and then asynchronously begin backoff later due to the arrival of a new frame. Furthermore, it is assumed that the propagation delay is ignored. Therefore, all stations are actually synchronized to operate in slotted time and they can correctly sense the channel status at the beginning of a slot. As a result, the transmission of a frame only happens at the beginning of a time slot, and the start time of the transmissions of all frames is synchronous.

In the analytical models, an idle time slot is usually assumed to have a fixed duration of δ. On the other hand, a busy time slot may contain a successful frame transaction or a collision of multiple frames. Let $\Delta_{s,l}$ denote the duration of a busy time slot where a frame from the lth AC is successfully delivered. For a time slot with collision, its duration is determined by the longest transmission time among the collided frames. Hence, the slot duration is denoted by $\Delta_{c,l}$ where l is the priority whose frame length is the largest. The duration of a busy time slot is calculated by

$$\begin{cases} \Delta_{s,l} = AIFS_l + T_{PHY} + T_{MAC} + T_{DATA}\left(E[Y_l]\right) + SIFS + T_{ACK}, \\ \Delta_{c,l} = AIFS_l + T_{PHY} + T_{MAC} + T_{DATA}\left(E[Y_l]\right) + EIFS, \end{cases} \quad (3.8)$$

where $E[Y_l]$ is the average payload length of the traffic with the lth priority, T_{PHY}, T_{MAC}, T_{DATA}, and T_{ACK} are the average transmission time of the PHY header, MAC header, payload, and ACK frame, respectively.

(4) Channel Access Zones in Prioritized MAC

As mentioned in Sect. 3.1.2, different AIFS intervals can be employed for ACs in the prioritized contention-based MAC. A station starts/resumes counting down its backoff counter after the channel is idle for the corresponding AIFS. An AC with a higher-priority is assigned with a smaller value of AIFS so that the traffic flows of this AC can reduce backoff counters and in turn access channel earlier than the lower-priority flows. Consequently, channel access in some time slots is restricted to a subset of flows with high priority only, and transmission/collision probabilities will vary in different time slots.

Figure 3.16 illustrates the access zones of two ACs (AC2 has a higher priority than AC1). Suppose that the AIFSs of the two ACs are $AIFS[1]$ and $AIFS[2]$ ($AIFS[2] < AIFS[1]$). The time period between two adjacent busy time slots except the AIFS for the higher-priority AC, $AIFS[2]$, is divided into two contention zones. In Zone 1, only AC2 stations can access the channel because the channel has been idle for $AIFS[2]$ but not for $AIFS[1]$ (the AC1 stations are still waiting). The AC1 stations are allowed to transmit only in Zone 2 where all the stations belonging to both AC1 and AC2 contend for channel access. Consequently, the competing stations will confront different contention levels in these contention zones. Please note that Fig. 3.16 can be extended to accommodate more ACs with different AIFS values and contentions zones (see Fig. 1 in [60]).

(5) Fixed-Point Analysis of Channel Access

The fixed-point is an important concept in functional analysis [148] which has been widely applied to determining the existence and uniqueness of solutions by pure mathematicians. Fixed-point theorems give the conditions under which variable mappings (by single or multi-valued mathematical functions) have solutions. Over the last several decades, it has been found that the fixed-point theory is a very useful and powerful tool in studying nonlinear systems. A diversity of fields such as biology, chemistry, economics, engineering, game theory, and physics have employed the fixed-point techniques. A well-known application of the Banach fixed-point theorem is the Newton's method proposed for nonlinear algebra equations.

The system using the traditional contention-based MAC is homogeneous because all stations (flows) adopt the same backoff parameters (like IEEE 802.11 DCF). Due to the symmetry, stations will have equal transmission probabilities and collision probabilities in their backoff processes. Brouwer's fixed-point theorem guarantees

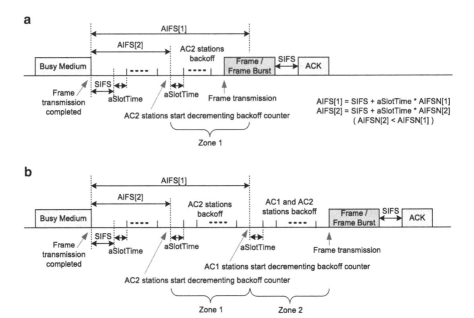

Fig. 3.16 An illustration of the channel access zones for prioritized ACs. (**a**) One or more AC2 stations transmit in Zone 1. (**b**) One or more AC1 and/or AC2 stations transmit in Zone 2

that there exists a balanced, unique fixed-point solution for the behaviors of the competing stations in a network [57]. Then the collision probabilities of all stations correspond to the fixed point. Similarly, in a heterogeneous WLAN using the prioritized MAC (like IEEE 802.11e), the flows belonging to the same AC are still equivalent and should have common backoff behaviors in the steady state. It has been shown that under certain conditions, the uniqueness of the fixed point for both the homogeneous and heterogeneous networks can be ensured [108]. Then the saturation throughput of the network can be calculated from the channel access statistics of the stations at this unique fixed point.

As mentioned above, a WLAN using the contention-based MAC is a fixed-point system. In the literature, the equilibrium channel access behavior of stations has been analyzed and the existence of the fixed-point solution is discussed [127]. Let **p** be the vector of the collision probabilities of all the stations in a WLAN. The collision probability of the nth station, p_n, is determined by the transmission behaviors of the other stations in a time slot, and is given by

$$p_n = H_n(\tau_1, \tau_2, \ldots, \tau_N) = 1 - \prod_{n'=1, n' \neq n}^{N} (1 - \tau_{n'}), \qquad n = 1, 2, \cdots, N, \qquad (3.9)$$

where $\tau_{n'}$ is the transmission probability of the n'th station.

When the system is in the equilibrium (steady) state, the parameters of the backoff behaviors of the stations should satisfy the following equation set

$$p_n = H_n(\tau_1, \tau_2, \ldots, \tau_N) = H_n(G_1(p_1), G_2(p_2), \cdots, G_N(p_N)), \qquad n = 1, 2, \cdots, N, \tag{3.10}$$

where $\tau_n = G_n(\cdot)$ is the function depending on the variable of p_n. The N equations in (3.10) can be written in the compact form of vectors as

$$\mathbf{p} = \mathbf{H}(\mathbf{G}(\mathbf{p})), \tag{3.11}$$

which is the multidimensional fixed-point equation.

As we can see, the independent variables in (3.11), p_n, are probabilities and thus in the range of $[0, 1]$. Hence $\mathbf{H}(\mathbf{G}(\mathbf{p}))$ in (3.11) [or the equation set in (3.10)] defines a continuous mapping from $[0, 1]^N$ to $[0, 1]^N$. According to Brouwer's fixed-point theorem, there exists a fixed point in $[0, 1]^N$ which satisfies the equation in (3.11).

Equation (3.9) is the nth component of the fixed-point equation, and can be written as

$$(1 - p_n) = \prod_{n'=1, n' \neq n}^{N} [1 - G_{n'}(p_{n'})], \qquad n = 1, 2, \cdots, N. \tag{3.12}$$

We can multiply both sides by $[1 - G_n(p_n)]$ and get

$$(1 - p_n)[1 - G_n(p_n)] = \prod_{n'=1}^{N} [1 - G_{n'}(p_{n'})], \qquad n = 1, 2, \cdots, N. \tag{3.13}$$

We can see that the right-hand side of (3.13) is independent of n. Therefore, the solution of the fixed-point equation in (3.11) should satisfy

$$(1 - p_i)[1 - G_i(p_i)] = (1 - p_j)[1 - G_j(p_j)], \qquad i, j = 1, 2, \cdots, N. \tag{3.14}$$

Equation (3.14) has provided a *necessary condition* for the solutions of (3.11).

In a homogeneous network using DCF, we have $p_i = p_j$ for all $1 \leq i, j \leq N$ due to the fairness of the protocol. The fixed point of a homogeneous IEEE 802.11 WLAN is unique and also balanced. Under certain conditions (safe ranges of node backoff parameters), a heterogeneous network using IEEE 802.11e EDCA has a unique fixed point and obviously the solution of $\mathbf{p}$ is an unbalanced fixed point. In these cases, the unique fixed point can capture the long-run average channel access operations of the stations and accurately predict the saturation network throughput [108].

The specific fixed-point equations can be derived using different approaches/models, as will be discussed in the following subsection. In a practical network, the parameters and statistics of the channel access behaviors should be identical for the stations belonging to the same AC, and otherwise for different ACs. In the numerical solutions of the fixed-point equations, the results of only one representative station of each AC need to be calculated. The numerical computations are repeated for the representative stations of the ACs. Since the number of ACs is much smaller than that of the stations, the computational complexity can be reduced considerably.

(6) Throughput

The aforementioned design principles in developing analytical models are unified for both saturated (every station always has backlogged frames to transmit) and unsaturated networks. It is worthy to notice that the calculation of throughput is different for the two cases. For a saturated network, the per station normalized throughput is just the ratio of the frame payload length (e.g., in bits) over the frame service time (or end-to-end delay). This is because one frame is successfully delivered in a renewal cycle. Therefore, the main task of the analytical models is to obtain the average frame service time for each traffic class. The per-station throughput of the nth station, denoted by η_n, is

$$
\begin{aligned}
\eta_n &= \frac{E[\text{payload delivered in a renewal cycle}]}{E[\text{duration of a renewal cycle}]} \\
&= \frac{E[\text{payload delivered in a time slot}]}{E[\text{duration of a time slot}]}.
\end{aligned} \tag{3.15}
$$

On the other hand, in an unsaturated network, a station may be idle without frames to transmit. Thus, the per station throughput should be the traffic arrival rate minus the frame dropping rate. Then, the delay performance can be obtained from the throughput results via the Little's law. We need to pay particular attention when applying the Little's law for delay computation due to the retry limits [14].

(7) Frame Service Time

The frame service time is defined as the time needed to successfully deliver the head-of-queue frame of a station to its target station, i.e., the time interval from the moment when the frame begins to perform backoff (i.e., becomes the head-of-queue frame in a buffer) to when the ACK for this frame is received. For a saturated station, it is also the time interval between two consecutive successful frame transactions. Please note that the average service time includes the possible transmission failures caused by collisions, i.e., dropping frames due to exceeding the retry limit.

According to this definition, the average frame service time of the nth station, denoted by ζ_n, can be calculated as

$$
\zeta_n = \frac{E[\text{average payload length}]}{\text{per-station throughput}}. \tag{3.16}
$$

(8) Network Model Assumptions

According to the general protocol models for contention-based channel access, the following conditions and behaviors about the random access are usually assumed as a starting point in all the three analytical models.

- The number of stations (terminals, nodes, or flows) is finite, and they are in a single-hop coverage (multi-hop is not considered as it is operated in the network layer). In a single-cell WLAN, the stations contend with each other for the channel access opportunities.
- The analytical models are usually developed under the condition that the channels are assumed ideal. The channel transmission error and capture effect are not considered. In other words, the frame transaction failures are only due to collisions caused by simultaneous transmissions from multiple stations. The frame reception failures caused by imperfect channels can be embedded in the analytical models, for example, following the approach presented in [159].
- The hidden terminal problem is not considered as the stations are all in a single-hop coverage of each other (i.e., a single collision domain). The propagation delay is ignored (assumed to be zero).
- Only one data frame is transmitted per TXOP. Consequently, the duration of the successful transmission time slot is a constant, including the corresponding AIFS, average payload transmission time (with average payload length and fixed data rate), physical and link-layer overhead, ACK frame, and SIFS interval, as given in (3.8). The time for control message (such as RTS/CTS) exchange should also be counted if applicable.
- The collision time slot is a constant. In the case of CSMA/CA using RTS/CTS, the collision time is RTS plus CTS. Otherwise, without RTS/CTS exchange, the collision time is given in (3.8) [12, 127]. For calculation simplicity, the duration of a collision time slot is sometimes approximated by that of a successful transmission slot.
- For ease of understanding, the models are presented assuming that each station has only a single queue of one AC. As presented in Sect. 2.1, IEEE 802.11e EDCA allows multiple separated queues for different ACs at one station. Each queue is used for one AC and behaves as a single contending entity. When the backoff counters of two or more queues reach zero, a virtual (internal) collision occurs. According to the rules of EDCA, the frame from the queue with the highest priority among the colliding ones is selected and transmitted on the radio channel. Meanwhile, the lower-priority queues will perform exponential backoff. For convenience, assume that each station carries only the traffic of a single AC and thus the virtual collision is not considered. The analysis for the more general case where a station has multiple queues for different ACs is discussed in [108].

3.3.2 Classic Analytical Frameworks and Models

(1) Markov-Chain Models

The seminal work by Bianchi [12] modeled the binary exponential backoff mechanism of a single station as specified in the IEEE 802.11 DCF protocol using a two-dimensional discrete-time Markov chain. The key principle is the "decoupling" approach. The stations are considered to be independent and statistically behave the same in channel access contention due to the fairness of the MAC protocols, as discussed in Sect. 3.3.1. A Markov chain is designed to model the backoff evolution process of an individual station, labeled as the "tagged" station, instead of jointly modeling the operations of all the competing stations. With the help of the Markov chain, we can write a set of nonlinear fixed-point equations which can be solved using numerical methods. By using the Markov chain, the transmitting probability which is the key parameter of a station's operation, is derived. Then the other parameters such as the collision probability are obtained. It has been shown that such a decoupling approach is valid to predict the statistical performance of the competing stations.

However, there are a couple of limitations in Bianchi's model. The limit of the retransmission times and the maximum backoff CW size which have been defined in the standard are not considered. But this work opened up a new horizon in exploring the contention-based MAC protocols theoretically. Numerous further research following the approach has emerged and tried to describe the standards more precisely. The retransmission limit is included in the models proposed by Wu [139] and Chatzimisios [20]. In their models, a frame will be dropped if the number of retransmissions reaches the pre-determined limit. Zhang [153] and Xiao [143] improved the model by introducing the probability that the backoff counter may be frozen due to the channel status, as specified in the standard. All of these models are designed to describe the channel access behavior under the saturation situation, i.e., a station always has backlogged frames to transmit.

For the unsaturated network scenario, there are mainly three approaches adopted to analyze the MAC protocols in the literature. Engelstad [38] and Malone [67] added new channel states in the Markov model for a station which correspond to the idle time without any frame to send. Following another approach, Zhai [149] and Tickoo [126] introduced the new parameter called queue utilization which was the probability of at least one frame backlogged in the buffer. Then the transmission probability of a station in the saturation-station model was scaled by the queue utilization to describe the unsaturation situation. The third approach models unsaturated networks by considering the number of stations which had backlogged frames and were currently actively contending for the channel access, instead of considering all the stations in the network (i.e., the stations that did not have frames ready for transmission were not counted). Thus the unsaturated networks could be analyzed. Such analytical models can be found in [41, 137] and [43]. In the multi-dimensional Markov chain proposed by Garetto and Chiasserini [43], the channel states included not only the backoff process but also the number of frames in the queues and the total number of active stations in the network.

Since the prioritized contention-based MAC is considered as an extension of the traditional protocol, most of the analytical models for the prioritized MAC (such as IEEE 802.11e EDCA) are based on the modifications of those for the DCF, by modifying or extending Bianchi's model to incorporate the differentiation of the AIFS and/or CW size. Some works have considered only the effects of the differentiated CW size, such as [48, 105, 141] and [53]. The models proposed in [142] and [145] have evaluated the differentiation effects of not only the CW size but also AIFS. To accommodate the prioritized channel access behavior, the authors in [160] and [142] changed the transition rates in the original Markov chain. The original bi-dimensional Markov chain was extended to tri-dimensional in [8, 158] and [161], and it was even extended to multi-dimensional in [145]. Zhao et al. [158] proposed a new Markov model to incorporate both the different CW sizes and AIFSs. To model the effect of AIFS, the state-space of the Markov chain is extended to include the number of slots which have elapsed since the last transmission activity in the channel. Bianchi et al. have also proposed a new analytical model for the prioritized MAC based on the differentiated AIFS which has high complexity [13]. Considering that only stations of certain ACs can attempt to transmit frames in certain network states, Robinson et al. [110] proposed a new approach which modeled the evolution of the differentiated channel states as a Markov chain. The transition probabilities were obtained based on the transmission attempt probabilities of the stations. The proposed analysis led to a fixed-point formulation of the prioritized channel access process.

(2) Mean Value Analysis

The mean value analysis is another analytical technique widely accepted for theoretically modeling the contention-based MAC protocols. The Markov-chain models focus on the microscopic behaviors of the contending stations, such as the detailed change/update of the backoff counter values over each generic time slot and the transitions of the backoff stages. However, in order to capture the differentiated channel access behaviors, the high dimensionality of the Markov models results in high complexity.

Compared with the Markov-chain based method, the channel access contention is evaluated at a higher level in the mean value analysis [18, 59, 61]. The mean value method analyzes the macroscopic behavior of a competing station through the channel access cycles and the long-run average performance. The mean value analysis directly describes the equilibrium characteristics of stations in the long-run channel access behavior, and then the average steady-state statistics are computed through the fixed-point iteration. Similar to the Markov-chain based approach, it is assumed that the probability that a station initiates a transmission in a slot is constant in its backoff process. Since a station regenerates the same channel access procedure for every new MAC frame, the complete service periods for frame transmissions form renewal cycles, which leads to a renewal process. Consequently, the backoff and frame transmission procedure can be studied based on a whole frame service cycle, and the average length of a renewal cycle is the average frame service time. From the renewal reward theorem, the transmission opportunities of a station can be regarded as the rewards during the renewal cycles [24, 59].

The mean value analysis can be employed to study both the traditional non-prioritized MAC protocols and the differentiated QoS-oriented MAC protocols for both saturated and unsaturated networks. Kumar et al. [57] significantly simplified and generalized the Bianchi's model [12] for the IEEE 802.11 backoff mechanism. In this work, the authors employed a one-dimensional fixed-point equation set for the collision probability experienced by the traffic flows using DCF. To consider unsaturated traffic and unbalanced traffic load often seen in a realistic infrastructure-based WLAN, the mean value analysis of the IEEE 802.11 DCF has been developed by Cai et al. [18]. The authors have obtained the conditional collision probabilities and frame service rates of the access point (AP) and stations, respectively, and the queue utilization ratio.

For prioritized MAC protocols, Ling et al. [60] further generalized the mean value analysis framework to model the EDCA protocol in the steady state. The system performance metrics such as the mean frame service time and the normalized station throughput were obtained. Using this method, Liu et al. [63] studied the performance of the prioritized channel access (PCA, defined in WPAN WiMedia-368 standard, see Sect. 2.2) protocol. This work considered two user traffic classes. The traffic of voice/video streaming and other multimedia applications has a higher priority in channel access than the background data traffic such as file transfer. The frame backoff and transmission procedure of a tagged station in either class was modeled, and the probability generating function (PGF) of the frame service time was obtained. Thus, the statistics of the system performance metrics could be calculated. In order to analyze the differentiated channel access mechanism, Ramaiyan et al. [108] extended the fixed-point analysis method for heterogeneous networks, using the *multidimensional fixed-point equations*.

(3) Backoff Counter Distribution Analysis

The Markov-chain and the mean value analysis study the network performance through per-slot statistics, such as the transmission and the collision probabilities defined with respect to a generic time slot. An alternative modeling technology, instead of relying on the per-slot transmission/collision probabilities, is to characterize each competing station through the steady-state distribution of the random backoff counter after one or multiple simultaneous transmission trials in the network.

The value of the nth station's backoff counter is a discrete-time discrete-state random process, denoted by $B_n(t)$ where t is the index of the generic time slots of the network. The activities of the wireless channel can be described in cycles and each cycle is composed of an initial random waiting time (all stations are deferring due to the contention avoidance) and a transmission slot where a frame may be successfully delivered (if only one station transmits in the slot) or a collision happens (if two or more stations transmit in the slot). The time instant when a packet transaction or a collision finishes is the start of the next channel access cycle.

Let $\xi = 1, 2, 3, \cdots$ indicate the start of the channel access cycles. At the beginning of a cycle, i.e., the time instant ξ, the nth station ($n = 1, 2, \cdots, N$) begins to compete for the channel access and its operation in the current cycle depends on its constant AIFS and the initial value of its backoff counter at the start of the cycle which is denoted by $b_n(\xi)$. Obviously, the initial values $b_n(\xi)$ of all the stations ($n = 1, 2, \cdots, N$) determine the number of idle slots before the transmission event and the transmission result (successful transaction or collision) at the end of the cycle. The model is observed at the beginning of each contention cycle, i.e., the discrete-time immediately before the expiration of the minimal AIFS after a channel busy slot (transmission or collision event) in the network.

Due to the distributed nature and fairness of the contention-based MAC schemes, two assumptions can be accepted. First, similar to the Markov-chain approach, the decoupling approximation is assumed. At the time instant ξ, the statistical behaviors of the competing stations are independent from each other, and the statistics of the channel access behavior of a station is determined by the distribution of its backoff counter process at ξ, i.e., $b_n(\xi)$. Second, the statistical distribution of $b_n(\xi)$ seen by a station at the beginning of a cycle is independent from the specific cycle. In other words, the statistics of the channel access behavior of a station (i.e., backoff counter evolution process) are stationary in different cycles. This also results in the fact that the cycles are independent and the channel activity statistics are regenerated in every cycle. This principle is actually similar with the reward renewal cycle in the mean value analysis discussed above. Based on these approximations, a WLAN using the contention-based MAC is a fixed-point system that can be solved through numerical techniques. The fixed-point solution is the distributions of the backoff counters at the beginning of each cycle. The throughput and other performance metrics are derived based on the statistics of the renewal cycles, as discussed in Sect. 3.3.1. Readers are referred to [127] for the detailed derivation of the backoff counter distribution and the performance metrics.

3.4 Summary

We focus on the distributed, contention-based MAC mechanisms in this chapter, discussing the design, performance evaluation, and analytical models of both the traditional and prioritized protocols. We have first investigated in detail the traditional and the prioritized channel access mechanisms. The principles and schemes in the frame transmission scheduling and resource management are discussed. Furthermore, to evaluate the parameter differentiation mechanisms, the ad hoc networks using the prioritized contention-based MAC are simulated by OPNET. The performance of the end-to-end frame delay and throughput of ACs with different AIFSs and CW sizes are presented.

To design QoS-aware resource management mechanisms in wireless networks, the models and analysis of the MAC schemes are needed which can accurately predict the performance for a wide range of parameters in single-hop network scenarios. The models for the prioritized MAC are mostly originated in those proposed for the traditional MAC (such as the 802.11 DCF), and extend the original models by considering differentiated backoff parameters such as AIFS and CW size. In this chapter, we have discussed in detail three kinds of analytical approaches.

The Markov-chain model for the prioritized contention-based medium access (e.g., EDCA and PCA) is first introduced. This approach focuses on the microscopic behaviors such as transitions between backoff counter values and stages, and thus usually leads to high accuracy and complexity. In addition, the fixed-point theorem arising from the backoff process of the distributed channel access process is discussed, such as the uniqueness and balance of the fixed point in homogeneous and heterogeneous networks.

The mean value analysis is simple yet effective for the distributed prioritized MAC of WLANs. Instead of studying the details of the stochastic process using a Markov chain, this approach tries to approximate the average values of the system variables, such as the average number of transmission trials and the average frame service time. This technique utilizes the closed-form expressions for the transmission probability, collision probability, and throughput, thus facilitating the analysis of the system performance based on various parameters. It has been shown that even though the mean value analysis omits many system details, it still achieves good accuracy in performance prediction [18, 59–61, 63].

The first two modeling approaches are both based on the "per-slot" statistics, i.e., transmission and collision probabilities in every generic slot. The variables are computed by fixed-point iteration for the channel access operation in slots. Different from the first two modeling methodologies, the third one does not rely on the per-slot transmission/collision probabilities, but analyzes the network performance by characterizing the steady-state distributions of the random backoff counters of the competing stations. The backoff counter distributions observed after transmission trials are calculated. Then the channel access performance metrics in terms of throughput and delay as well as the low-level statistics such as the backoff counter distribution can be obtained with high accuracy.

Obviously, a tradeoff has always been made between model accuracy and complexity. Under the conditions of certain contention parameter settings, the correlations among consecutive idle slots and among consecutive channel accesses can be neglected and the approach of the persistence models is pretty accurate. Some research works preferred to modeling the channel access processes with simplification, in order to facilitate the cross-layer design and theoretical analysis. For example, the MAC models can be incorporated with the admission control algorithms, transport protocols, and physical-layer channel models [22].

Chapter 4
Resource Reservation

Distributed resource reservation has attracted great interest from academia, industry, and standards organizations thanks to its advantages in guaranteed QoS provisioning. Given the shared wireless media, contention-free protocols rely on additional coordination among competing stations. With the bandwidth and delay requirements, multimedia traffic often needs to reserve certain channel time. There are two kinds of contention-free protocols. One is the centralized allocation, such as IEEE 802.11 point coordination function (PCF) and IEEE 802.15 channel time allocation (CTA), where one station is designated or elected as the piconet controller and manages channel allocation. The other is distributed reservation protocols, such as the WiMedia ultra-wideband (UWB) MAC, where each station exchanges the channel availability information with neighbors through the broadcast beacon messages in a well-defined superframe structure and negotiates with the destination for channel allocation. The latter has advantages in scalability and fault-tolerance. In this chapter, we show two reservation algorithms, *subframe-fit* and *isozone-fit*, initially specified in [49] for the distributed reservation mechanisms. The analytical models proposed in [31] are presented and validated by simulations using NS-2 and an MPEG-4 traffic generator.

4.1 Introduction

4.1.1 Channel Reservation Principles

(1) Reservation-Based Transmission Mechanisms

In the previous chapters, we have seen that multiple access mechanisms (i.e., MAC protocols) are critical to coordinate a number of wireless stations for resource management and scheduling. Since generally wireless/mobile users are randomly

© The Author(s) 2017
R. Zhang et al., *Resource Management for Multimedia Services in High Data Rate Wireless Networks*, SpringerBriefs in Electrical and Computer Engineering,
DOI 10.1007/978-1-4939-6719-3_4

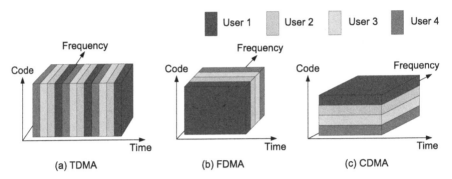

Fig. 4.1 Typical channel resource reservation schemes in the time, frequency, and code domains

distributed inside a radio coverage area and share a common channel, they are required to compete for opportunities to transmit their data. In contention-based random access mechanisms, such as ALOHA and CSMA discussed in Chap. 3, the channel access is controlled in a distributed manner and thus is quite flexible: each user decides when to send its frames individually and independently. Nevertheless, frame collisions among users can happen frequently which cause throughput and service time variation. In particular, when traffic load is heavy, the contention-based schemes suffer from serious collisions due to the severe contention, resulting in drastically decreased throughput and increased delay. To guarantee bandwidth and delay, real-time voice/video traffic often needs to reserve certain channel time for contention-free transmission. Despite the contention nature of wireless media, channel reservation can be done by additional coordination among competing stations, and it can provide more stable throughput and delay. In the reservation-based resource management, each user is pre-allocated an exclusive portion of channel resources, such as bandwidth, time, or codes, and it can transmit frames using the assigned resources only. The typical channel resource reservation in the time, spectrum, and code domains are time division multiple access (TDMA), frequency division multiple access (FDMA), and code division multiple access (CDMA). The principles of the three reservation mechanisms are illustrated in Fig. 4.1.

Combined with packet-based multiple access, the resource management by reservation can effectively support QoS for integrated multimedia and data traffic where several different ACs exist with various throughput and delay requirements. For example, Cheng et al. [23] have investigated a wireless clustered mesh network with an orthogonal frequency division multiplexing (OFDM) physical layer. In the multi-channel environment within a cluster, by allocating subcarriers to different stations, simultaneous transmissions are allowed and thus the network capacity is increased. Meanwhile, subcarrier reservation and collision-free packet scheduling can enable QoS provisioning for multimedia traffic. In addition, power allocation for the subcarriers needs to be considered due to different channel fading characteristics among the subcarriers. Therefore, a joint power-subcarrier-time resource allocation is studied. The authors proposed an efficient intra-cluster packet-level resource allocation approach with QoS provisioning, named Combined-KKT-GA. The proposed approach combines the merits of a Karush-Kuhn-Tucker (KKT)-driven approach (low complexity) and a genetic algorithm (GA)-based approach (near-optimal performance). This approach has jointly considered power allocation, subcarrier allocation, packet scheduling, and QoS support in resource management in a wireless network using OFDM in the PHY. It has been shown that the proposed approach has good throughput and packet dropping rate performance [23]. In [132], Wang et al. proposed a collision-free MAC scheme for a single-channel wireless mesh backbone to support multimedia applications. By taking the fixed network topology into account, collision-free transmissions are scheduled in a deterministic way in a TDMA manner. The channel time is slotted and one slot consists of two portions: control part and transmission part. The former is further divided into several mini-slots used for time allocation, i.e., to determine whether or not a router can transmit its frames in that slot. The mini-slot assignment algorithm, spatial reuse scheme, and congestion control mechanism have been designed. The proposed scheme can provide guaranteed priority access for real-time traffic and, at the same time, ensure fairness for data traffic. Jiang et al. [55] have proposed an interference-aware distributed MAC scheme for a CDMA-based wireless mesh backbone considering the unique networking characteristics. Taking into account the fixed locations of wireless routers, the time slot/power/rate are allocated based on the transmission distances and interference levels, and the admission of new calls is determined by the maximum sustainable interference. Fine-grained QoS can be achieved by the designed resource reservation scheme.

On the other hand, reservation-based channel access also faces many challenges. For example, by using TDMA, user terminals have a $1/N$ duty cycle where N is the number of users. The signals will have an aperiodically pulsating power envelope. This presents a challenge to designers of portable RF units. Also, the bit rate needs to be increased by N times. Consequently, equalization against multipath is generally needed to mitigate the inter-symbol-interference (ISI). Meanwhile, strict synchronization among the stations is required which is non-trivial in practical implementation. Furthermore, time slot and frequency allocation and management entail a certain extra complexity in TDMA and FDMA systems, respectively.

In this chapter, since TDMA has been widely adopted in digital cellular networks and personal communication systems, we focus on the resource management in TDMA. Reader can refer to the extensive literatures for that in other reservation-based multiple access schemes such as FDMA and CDMA. TDMA has been used in the European, North American, and Japanese second-generation digital cellular systems (GSM, IS-54, and PDC, respectively) [46, 103], and in several wireless personal communications service (PCS) systems including the European digital cordless system (DECT [46]), the Japanese personal wireless system (PHS [104]), and the wireless access system proposed by Bellcore (WACS [44] or PACS).

In a TDMA wireless network, multiple users share a common carrier frequency band in the time domain to communicate with a BS or each other. Each user is allocated one or more time slots within a channel time frame and transmits only in its own assigned time slot(s). Inter-user interference is prevented by strict adherence to time slot schedules, and by guard times and time-alignment procedures which prevent overlaps even with different propagation times.

A channel time scheduling frame (usually called "*superframe*") generally contains a number of time slots which are the units for time allocation. A time slot can accommodate user data bits (perhaps including channel coding bits for error detection and correction) and extra bits for synchronization, adaptation, control, guard time, etc. The smaller the fraction of the frame devoted to these "overhead" bits, the more efficient is the TDMA frame design. For instance, in the GSM, PHS, and DECT systems, about 30 % of the total transmitted bits are for overhead. The IS-54 and PDC systems, on the other hand, incorporate roughly 20 % extra data bits for overhead. Most of the overhead in the GSM and IS-54 systems is the training sequences for the adaptive equalizer, while in DECT the major portion of the overhead is used for system control [39]. For a given amount of overhead, higher efficiencies could in principle be achieved by increasing the time slot duration (to reduce the portion of the overhead in the total data rates or channel time). However this may have the adverse effects of increasing the total transmission delay for time-sensitive traffic and hampering the system's ability to adapt to rapid changes in the fading environments. Goodman [46] provides a comprehensive description of the frame structures of several TDMA systems.

TDMA has several advantages compared with other reservation-based multiple access techniques. Working at the same carrier frequency, the networking devices can employ the common radio frequency (RF) and modulation/demodulation equipment. Thus the complexity and cost for the hardware systems are reduced significantly. Another advantage is that, by allocating more or fewer time slots to each individual station, it is convenient to adjust the bit rates for users according to their instant traffic load and QoS requirements. This is especially desirable for supporting integrated services. Compared with CDMA, TDMA does not require strict power control for the near-far effect, since inter-user interference is avoided by using alternative time slots instead of by processing gain obtained from spectrum spreading.

(2) MAC Protocols for Channel Reservation

Polling and reservation are two strategies for contention-free MAC protocols, and the latter can be more efficient and are often used to provide QoS guarantees for delay-sensitive traffic over wireless links.

By using reservation-based MAC protocols, priorities in channel access and resource utilization can be conveniently provided in the reservation allocation stage. Several resource allocation schemes have been designed. The resource management strategy suggested in some protocols is that all low-priority flows should refrain/backoff from transmission until all higher-priority flows have already successfully accessed the channel [72, 121]. Some other protocols, however, adopt the policies which are less restrictive to low-priority flows [27, 58]. According to these schemes, low-priority flows are allowed to contend for channel access against high-priority flows. But, the low-priority flows are required to wait for longer time on average in order to protect the transmissions of the delay-sensitive flows. The idea is similar to that of the prioritized contention-based MAC protocols discussed in Chap. 3. In recent years, a number of MAC protocols for resource-reservation channel access have been designed, such as TDD ALOHA Reservation [123], DR-TDMA [42], RAMA [4], R-ALOHA [66], PRMA [36], and DRMA [106].

Most of the MAC protocols for TDMA focused on centralized schemes with a controller [34, 50]. Traditional contention-free protocols such as IEEE 802.11 point coordination function (PCF) and IEEE 802.15 channel time allocation (CTA) normally follow a centralized design, e.g., one station is designated or elected as the piconet controller, managing channel allocation in its vicinity. A user who needs to send data transmits a request message/frame (usually by contention-based channel access) to the *controller/coordinator* for channel reservation. Upon the arrival of a new request, the controller will determine which slot is free and can be used in a TDMA manner. After successful reservation, the user is then able to send its data frames without contention. IEEE 802.15.3 [86] is one of the standards that target on enabling high date-rate multimedia applications in WPANs. In this standard, all communications between all devices can only be carried out through or enabled by the *piconet coordinator*. Such schemes have certain limitations as follows.

- All communications have to go through or be enabled by the controller; if the controller fails, the entire network may not work any more. In IEEE 802.15.3, failure of the coordinator would disable the entire piconet for a considerable amount of time before all other devices recognize the situation and re-elect a new coordinator [102].
- A centralized scheme does not scale well with a large or dynamic number of stations.
- These TDMA systems generally allocate one block of time slots which is often fixed to each flow/station per superframe. Even though this block could have a variable length, the interval between two consecutive transmission slots for one station is at least the duration of a superframe. Thus, it may not be flexible enough to meet the maximum delay requirement for QoS of delay-sensitive traffic.

The distributed resource reservation without central controller is more suitable for ad hoc networks. It has been advocated as a cost-effective approach to support high dynamic and integrated services in ad hoc networks. Among the distributed MAC protocols for resource allocation, one example is the distributed reservation protocol (DRP) specified in WiMedia UWB standard, ECMA-368 [49]. A well-defined superframe consists of many media access slots (MASs) and the slot allocation can be arbitrary. Each station exchanges the slot availability information with neighbors through the broadcast beacon messages in superframes and negotiates with the destination station for channel allocation. Such a distributed scheme has certain advantages in scalability and fault tolerance. It also allows stations to reserve multiple slots at different locations within one superframe to meet their bandwidth and delay requirements. Meanwhile, the unreserved slots are available for contention-based channel access.

There are also some research efforts on such distributed reservation protocols. On one hand, the performance of DRP with arbitrary slot reservation was analyzed. Wu et al. [140] have proposed a two-dimensional Markov chain to analyze the delay performance of DRP under a given reservation pattern. The model can only deal with hard reservation which is not suitable for delay-sensitive traffic with bursty arrivals. In [64], Liu et al. have investigated the performance of DRP in a more realistic scenario where the time slots are reserved non-uniformly in superframes and the communication link between a pair of transceivers is obstructed by a person (body shadowing). A cross-layer analytical model was proposed where the UWB multipath channel with shadowing was modeled by a finite-state discrete-time Markov chain. In addition to the time-varying channel state (channel fading), the reservation protocol and the induced reservation pattern in a superframe also affected the protocol performance. In order to analyze the frame delivery of a flow with arbitrary reservation pattern, the analytical model was developed based on the vacation queuing process where the interval between two reservation periods was regarded as the vacation period of the flow. The impact of the reservation pattern on the DRP performance was revealed in this work. Liu et al. [62] have further studied the impact on delay by different reservation patterns for UWB MAC taking into account the wireless shadowing channel. By cross-layer analysis, Zhang et al. [152] developed a general analytical framework which incorporates the UWB channel fading, joint error control (adaptive modulation and coding, auto-repeat request, and packet fragmentation), and arbitrary reservation pattern in superframes. Based on this model, the frame queueing process was analyzed and the link delay and loss performance was quantified. On the other hand, the reservation algorithms have been studied by Daneshi et al. in [29, 31] and [30]. The authors proposed two heuristic reservation algorithms to meet the application QoS requirements and ensure efficient channel utilization, which will be discussed in more details in the following sections.

4.1.2   Challenges and Issues

Resource allocation has been extensively studied in the field of operating systems for memory and disk management [134]. For memory allocation, paging-based approaches become predominant with fixed-size memory pages, and only the request size is considered due to the direct access of main memory devices. Although the location of disk blocks may affect the access delay, such difference is often ignored in disk management. In the context of distributed reservation protocols for wireless links, e.g., WiMedia UWB MAC, not only the size of the reservation, but also the tolerable delay should be considered. The constraints on where the reservation blocks can appear and how large they can be also introduce new challenges for reservation allocation algorithms.

To design broadband wireless multimedia communication systems using TDMA technologies, we must take the following requirements into account:

- accommodating dynamic variation of traffic,
- supporting QoS according to the traffic requirements,
- mitigating frequency/time-domain channel fading, and
- achieving high system capacity for more users.

In order to satisfy the above requirements and provide differentiated services, there are several challenges for reservation-based MAC protocols.

- *How to allocate resources efficiently:* Traffic flows do not always transmit at a constant rate and have considerable rate variation, so it is difficult to determine the amount of resources to reserve. If reservations are made at the peak rate to minimize the queuing delay, they leads to a significant waste of bandwidth and the system capacity is sacrificed much (for example, the number of supported video streams is reduced considerably). If reservations are made at the average data rate, considerable queuing delay and packet losses can occur during traffic bursts.
- *How to coordinate competing users:* As mentioned in Sect. 4.1.1, there are mainly two resource allocation mechanisms: the centralized one and the distributed one. The former is not quite flexible because a central controller is needed. The latter is more scalable but more complicated because stations need to negotiate with each other. Efficient and reliable reservation-based MAC is needed for the coordination in a wireless network, which is the prerequisite for resource reservation.
- *How to partition resources:* For example, in TDMA, all users need to be aligned with the time slots to start and finish transmissions. Thus synchronization among users is required which is non-trivial to realize in practice. In FDMA, different frequency bands are assigned to users, but the spectrum should be carefully allocated in order to avoid intolerable inter-user interference and also achieve high spectrum utilization.

In summary, to efficiently allocate time/bandwidth with minimum cost and optimal performance in reservation-based resource management is still a challenging issue and active research topic.

4.2 Distributed Channel Reservation Mechanisms

In this section, we present the work in [31] as an example of distributed reservation algorithms for QoS provisioning. The authors in [31] have discussed DRP, specified in the WiMedia ECMA-328 [49] for high-rate WPANs based on UWB. Each station exchanges the slot availability information with neighbors through the broadcast beacon messages in a well-defined superframe structure and negotiates with the destination node for channel allocation. Although the DRP in WiMedia UWB MAC is flexible, the standard only specifies the physical-layer and fairness constraints on possible allocations, not the reservation algorithms and their performance [69].

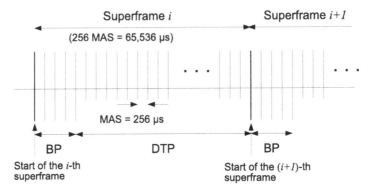

Fig. 4.2 The structure of the superframe defined in WiMedia ECMA-368

(1) Superframe Structure and Reservation Rules in ECMA-368

The ECMA-368 specified by WiMedia [69] is a fully distributed protocol for channel time reservation and has high flexibility. In this MAC scheme, the channel time is divided into fixed timeframes called *superframes*. The superframe duration is around $65,536\,\mu s$, including 256 medium access slots (MAS) with duration of $256\,\mu s$ each as shown in Fig. 4.2. A MAS is the time unit that a station uses for reservation. Each superframe has two main parts, a beacon period (BP) and a data transfer period (DTP). An availability information element (IE) is broadcast in a beacon message inside each BP. It indicates a station's current utilization of the MASs. There are two protocols supported during DTP: prioritized contention access (PCA) and distributed reservation protocol (DRP). PCA is similar to the EDCA in the IEEE 802.11e standard (discussed in Chap. 3), which provides differentiated channel access to wireless media with different priorities.

A two-dimensional structure of the WiMedia superframe has been proposed in [135] to elaborate the MAC policies. Each column of the 16×16 superframe matrix is called an *allocation zone*. For each column (zone), the transmission is in the row order. Allocation zones of a WiMedia superframe excluding the BP zones are grouped into four sets called *isozones*, i.e., isozone 0–3, as shown in Fig. 4.3. The MAS columns within the same isozone are distributed evenly across the superframe. More specifically, the MASs located in the same row and "adjacent" allocation zones within the same isozone are separated from each other by a fixed interval that depends on the isozone. Such an interval is referred to as the native service interval (SI) of the isozone.

WiMedia MAC standard has specified a set of rules for distributed channel reservation. These rules are defined to meet the physical-layer and fairness constraints. They limit both the size of the reservation blocks and their possible locations in a superframe: e.g., reservation requests should follow the order of isozones as much as possible. No single reservation block can have more than eight MASs, although a single reservation request can have multiple reservation blocks at different locations of the superframe. When a reservation block starts at row 4 and up to row 11,

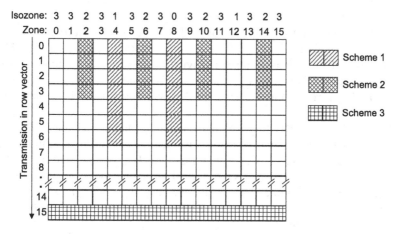

Fig. 4.3 A two-dimension view of the WiMedia UWB superframe [31]

the block cannot have more than four MASs; when starting before row 4 or after row 11, the block size can be larger or smaller than 4, respectively. For a particular reservation, row components (i.e., those containing reservation blocks in all columns) should be in the lower portion of the superframe matrix, and column components (i.e., all but row components) should be in the upper portion of the superframe as much as possible. Figure 4.3 shows three ways of reserving 16 MASs in the superframe. Compared with the reservation *Scheme 1*, *Scheme 2* has a smaller block size but is more evenly distributed. *Scheme 3* is a row component with blocks in all columns of the bottom row.

We need to create a reservation strategy for WiMedia MAC to assign MASs to flows based on their QoS requirements. Thus, we first need to determine how many slots are needed for a flow according to its delay requirement, which will be used by the reservation algorithm to find a feasible reservation. Upon a successful reservation, the owner will inform other UWB stations by updating the reservation bitmap in beacon messages. The reservation shall guarantee the service rate between the minimum and maximum ones requested by a flow and meet its maximum tolerable delay.

(2) Reservation Approaches

Making a reservation of network resources for video flows over WiMedia UWB channels requires the determination of the number of MASs needed and their locations within the superframe. The number of slots needed per superframe can be derived from the flow's traffic specification (TSpec). Delay requirement of each flow, specified in the TSpec, is defined as the maximum delay each packet of the flow can tolerate from the time it arrives at the MAC layer till it is transmitted. Satisfying the delay requirement for video traffic then depends on the service rate (due to the queuing delay) and the location of the reserved slots (due to service interval).

To obtain the number of required slots for a reservation, Daneshi et al. used the approach of equivalent bandwidth (EBW) in [31]. The EBW function characterizes the bit rate required to be reserved for a given flow with regard to its TSpec. More precisely, for a flow with a cumulative arriving function R over the time and for a fixed but arbitrary delay D, we can define the effective bandwidth $g = f(R, D)$ of the flow as the bit rate required to serve the flow in a work-conserving manner, with a delay bound of D.

There are different EBW functions in the literature, and in [31], the authors used a computationally simple approximation. The required bandwidth allocation, g, for a traffic source with a given overall flow loss ratio P_{loss} is suggested in [109] as: $g = a_1 * r + a_2 * \sigma^2/C$, where C is the channel capacity, r and σ^2 are the mean and variance of the traffic bit rate, respectively. We have $\sigma^2 = r(p - r)$ where p is the peak source rate. For the coefficients a_1 and a_2, [109] has suggested the following empirical approximation:

$$a_1 = 1 - 0.02 \log_{10} P_{loss},$$
$$a_2/a_1 = -6 \log_{10} P_{loss}. \tag{4.1}$$

In (4.1), P_{loss} can be chosen based on the acceptable packet loss ratio for the type of traffic under consideration. We here set $P_{loss} = 0.01$ for MPEG-4 video traffic with error concealment decoding. With this approximation for the service rate, a flow is guaranteed to be transmitted between its minimum and maximum requested data rate. Once g is obtained, the number of slots needed can be determined by the channel data rate and protocol overhead.

(3) Reservation Algorithms

Although the DRP in WiMedia MAC is very flexible, the standard only specifies the physical-layer and fairness constraints on possible allocations, not the reservation algorithms and their performance. As discussed above, the WiMedia superframe is treated as a two-dimensional matrix, and the maximum number of slots can be reserved is row/column-dependent. To make the tradeoff between allocatable slots and the intrinsic delay among them, we should model, analyze, and evaluate the two reservation algorithms, *subframe-fit* and *isozone-fit* proposed for distributed reservation in WiMedia MAC. Subframe-fit tries to follow the one-dimensional allocation and replication strategy by taking the reservation size and delay into account, while isozone-fit takes one step further to meet the WiMedia allocation constraints as well by following a two-dimensional allocation strategy. Thus, subframe-fit provides an optimistic bound for all WiMedia-compliant allocation schemes.

• Subframe-Fit Algorithm

The *subframe-fit* algorithm only considers the size and delay requirements of a flow. It does not take all the WiMedia MAC rules into account, which makes the algorithm computationally simple and provides an optimistic upper bound for achievable performance. Knowing how many time slots (denoted by m) a flow needs to reserve in addition to its delay requirement (denoted by d), the algorithm divides the superframe into subframes of length $C' = \lfloor C/k \rfloor$ time slots each, where C is the number of time slots in the superframe, $k = \lceil T/d \rceil$ and T is the superframe duration. Accordingly, the algorithm divides the request into subflows, each requesting $\lceil m/k \rceil$ time slots per subframe. Then it tries to fit each of these subflows in the appropriate subframe. A reservation can be made only if all of the subflows of a flow can be reserved.

Although subframe-fit is not fully compliant with the WiMedia MAC, its performance can be investigated as a baseline, since it provides an optimistic bound for all WiMedia-compliant algorithms such as isozone-fit. It also allows us to evaluate the performance penalty of WiMedia MAC policies to ensure fairness, and to investigate possible improvements of isozone-fit.

- **Isozone-Fit Algorithm**

With the number of slots required (m) and also the delay requirement (d), the isozone-fit algorithm starts from the isozone that has the closest natural service interval to the flow's tolerable delay (excluding the queuing delay) and searches through the superframe starting from that isozone for available slots. Isozone-fit tries to keep the superframe symmetric and well-structured, i.e., it allocates the blocks of the same flow in the same row of different columns. If a flow leaves and releases its reserved time slots, the superframe will still be well-structured. By keeping reservations symmetric, it may block some reservation requests, but future flows can fit easily and the superframe will have less fragmentation. This algorithm also follows the reservation rules of the WiMedia standard, which limits both the size of the reservation blocks and their possible locations in the superframe.

The basic isozone-fit algorithm only keeps one request in the current isozone, if the request can be accommodated; otherwise, the request will be dropped (or overflowed) to the next isozone with a higher isozone index, which normally has more available slots and tighter service interval and hence a better chance to accommodate the request. In the next section, we will further evaluate the case where one request can be accommodated in multiple isozones at the same time.

4.3 Modeling and Analysis

In this section, the analytical models for the ECMA MAC introduced in the previous section are presented. Since the reservation algorithms aim to support bursty traffic such as compressed video, we first specify how to handle traffic offered to the system. Then, the analytical models proposed in [31] for the two channel reservation algorithms, *subframe-fit* and *isozone-fit*, are presented.

(1) Traffic Decomposition

For both algorithms, each flow requires a certain number of time slots with a specific delay requirement. A request could either be carried (time slots reserved) or dropped. Suppose that there are n flows in the piconet and the ith flow request is characterized by (a_i, z_i, m_i, d_i) for the heterogeneous traffic class i ($i = 1, 2, \ldots, n$). The parameter $a_i = \lambda_i/\mu_i$ is the offered load where λ_i is the mean arrival rate and μ_i is the mean service rate. z_i is the peakedness of the flow, i.e., the variance to mean traffic intensity ratio that shows how much the offered traffic exceeds the system's actual capacity. Offered traffic is deemed peaky, regular (i.e., Poisson traffic), or smooth accordingly if z_i is greater than, equal to, or less than 1, respectively. m_i is the number of the requested time slots and d_i is the maximum tolerable delay of the flow according to Sect. 4.1.1.

In the models for these two algorithms, flows are modeled by the Beshai's two-component representation of the mixture of non-Poissonian multi-channel traffic [11] based on Delbrouck's approximation [32]. Delbrouck's solution replaced non-Poissonian flows by a series of Poissonian streams. Beshai's approximation only takes the first two components of this series into account and each subflow is approximated by an equivalent set of two Poissonian streams of m_i and $\zeta_i m_i$ slots per request [11], i.e., $\{(w_{i,1}, 1, m_i, d_i), (w_{i,2}, 1, \zeta_i m_i, d_i)\}$, where

$$w_{i,1} = a_i \times \frac{\zeta_i - z_i}{\zeta_i - 1},$$

$$w_{i,2} = a_i \times \frac{z_i - 1}{\zeta_i - 1}. \tag{4.2}$$

In the numerical evaluation of this representation, ζ_i is approximated by an integer greater than z_i.

(2) Subframe-Fit Model

To analyze the performance of the subframe-fit algorithm, the authors in [31] have proposed a model based on the replication idea of the algorithm, i.e., each flow is divided into subflows and the superframe is divided into subframes based on the delay requirements of flows. With this model, a request is carried when all its subrequests are carried by all subframes. Figure 4.4 shows the structure of this model. The model has a preprocessing unit, which takes the flows and applies the decomposition formulas above to each request for the replication model.

With the traffic decomposition, let $P_x = P(X = x)$ be the steady-state probability of exactly x simultaneously busy time slots in the system, and it is determined from [11] that

$$xP_{x,j} = \sum_{i=1}^{n} w_{i,1} P_{x-m_i} + w_{i,2} P_{x-\zeta_i m_i}, \qquad 0 < x \le C', \tag{4.3}$$

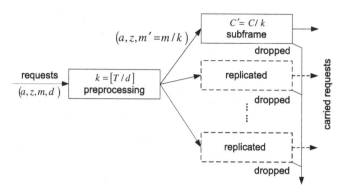

Fig. 4.4 The subframe-fit system model [31]

because $w_{i,1}$ and $w_{i,2}$ have the request size of m_i and $\zeta_i m_i$, respectively, and each slot of the subframe with C' slots is treated as a stream. The boundary and unit conditions for P_x are $P_x = 0$ for $x < 0$, and $\sum_{x=1}^{C'} P_x = 1$. The condition $x < 0$ is used for computation purposes only. For notation convenience, $Q_x = P(X > x) = 1 - P(X \leq x)$, $Q_{x-1} = Q_x + P_{x-1}$, $1 \leq x \leq C'$, and $Q_{C'} = P_{C'}$. Thus the blocking probability of each stream is given by

$$b_i = \frac{w_{i,1} Q_{C'-m_i+1} + w_{i,2} Q_{C'-\zeta_i m_i+1}}{a_i}, \tag{4.4}$$

i.e., for any new incoming request when any subframe is full, it will be dropped. Accordingly, the carried traffic of each stream is computed as xP_x, and the total carried traffic is

$$c = \sum_{x=0}^{C'} xP_x = \sum_{x=0}^{C'} x(Q_{x-1} - Q_x) = \sum_{x=0}^{C'} Q_x. \tag{4.5}$$

To evaluate the proposed algorithms, two system performance metrics are studied: blocking probability (β) and system utilization (u). Blocking probability shows the portion of the flows in the steady state that cannot reserve their requested time slots and therefore are dropped. System utilization shows the percentage of the superframe that is reserved in the steady state. Here, system utilization is computed by the ratio of the total traffic carried by the system to the total system capacity, i.e.,

$$u = c/C'. \tag{4.6}$$

Blocking probability of the system is computed as the sum of the blocking probability of each stream, i.e.,

$$\beta = \sum_{i=0}^{n} b_i. \tag{4.7}$$

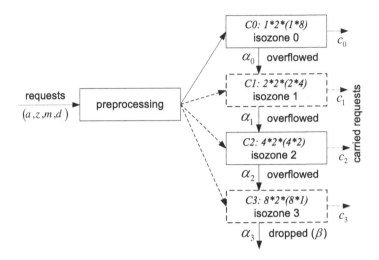

Fig. 4.5 The isozone-fit system model [31]

(3) Isozone-Fit Model

To evaluate the performance of isozone-fit, the WiMedia UWB superframe is modeled as a four-class non-Poissonian lossy server serving the flows [31], as illustrated in Fig. 4.5. This model uses the isozone structure of the superframe defined in the standard for class mapping; therefore each class represents one isozone of the superframe. For example, the first class (C_0) has one column that has two blocks of eight slots each for reservation, while the second class (C_1) has two columns and each has four blocks of four slots, available for four requests of eight slots. Because of this arrangement, each class of time slots has a natural service interval. This service interval is the minimum delay a flow would experience if it reserves time slots from a particular class.

A flow requests for a certain number of time slots with a specified delay requirement. This request is offered to an appropriate class based on the availability of the time slots and the flow's requested delay. If enough slots are available and they match the flow's requirement, they will be occupied for the duration of the flow in the piconet. Otherwise the request will be dropped to the next class. From the isozone structure, the next class always has a larger capacity and a smaller native service interval (SI); therefore there may be a feasible reservation for a flow dropped from the previous class.

To solve this model, Beshai's two-component representation of the mixture of non-Poissonian multi-channel traffic [11] is used again. The portion of the requests that is carried by each class is denoted as c_j ($j = 0, 1, 2, 3$) and the overflowed portion of them to the next class is denoted as α_j. The blocking probability is the portion of the requests that are dropped from the last class and it is defined as $\beta = \alpha_3$.

Let $P_{x,j} = P_j(X = x)$ be the steady-state probability of exactly x simultaneously busy time slots in class j and it is determined from [11] that

$$xP_{x,j} = \sum_{i=1}^{n} w_{i,1}P_{x-m_{i,j}} + w_{i,2}P_{x-\zeta_i m_i,j}, \qquad 0 < x \le C_j, \qquad (4.8)$$

where $P_{x,j} = 0$ for $x < 0$, and $\sum_{x=1}^{C_j} P_{x,j} = 1$. Again, the condition $x < 0$ is used for computation purposes. For notation convenience, $Q_{x,j} = P_j(X > x) = 1 - P_j(X \le x)$, $Q_{x-1,j} = Q_{x,j} + P_{x-1,j}$, $1 \le x \le C_j$ and $Q_{C_j,j} = P_{C_j,j}$. The blocking probability of each stream for each class is

$$b_{i,j} = \frac{w_{i,1}Q_{C_j-m_i+1,j} + w_{i,2}Q_{C_j-\zeta_i m_i+1,j}}{a_i}, \qquad (4.9)$$

if flow i is offered to class j; otherwise, $b_{i,j} = 0$. Thus, the total blocking probability of this class is computed as

$$\alpha_j = \sum_{i=0}^{n} b_{i,j}. \qquad (4.10)$$

Accordingly, the carried traffic for each class is

$$c_j = \sum_{x=0}^{C_j} xP_{x,j} = \sum_{x=0}^{C_j} x(Q_{x-1,j} - Q_{x,j}) = \sum_{x=0}^{C_j} Q_{x,j}. \qquad (4.11)$$

When the flow cannot be served by the initial class, the class with a higher isozone index can be used as an alternative for the overflowed traffic. Therefore, the traffic to the next class is no longer Poissonian. Wilkinson initially modeled the overflowed traffic based on equivalent random theory (ERT). For each class, if the blocking probability is greater than zero, the overflowed portion of the requests is offered to the next class. The ERT theory is to approximate the traffic by a Pascal (negative binomial) distribution. After that, Delbrouck [32] found the parameter similarities between Pascal and Bernoulli distributions and their convergence to Poisson distributions. It is then possible to use a unified approximation procedure to estimate the main congestion functions of peaky traffic. To determine the parameters of the overflowed traffic from a primary class to an alternative, the Wilkinson's approximation is utilized. In this model, which is based on the Riordan's formula [133], the moments of the marginal distribution are derived from the Erlang's B recursive formula as

$$E_x(a_{i,j}) = \frac{aE_{x-1}(a_{i,j})}{x + aE_{x-1}(a_{i,j})},$$

$$E_0(a_{i,j}) = 1, \qquad (4.12)$$

where $a_{i,j}$ is the offered load and x could have any value between 0 and the maximum server capacity C_j. Based on the Erlang'B formula, the approximation for the mean $(a'_{i,j})$ and peakedness $(z'_{i,j})$ of the overflowed portion is given in [133] as

$$a'_{i,j} = E_{C_j}(a_{i,j}) * a_{i,j} = b_{i,j}a_{i,j}, \qquad i = 1, 2, \cdots, n,$$

$$z'_{i,j} = 1 - a'_{i,j} + \frac{a_{i,j}}{C_j + 1 - a_{i,j} + a'_{i,j}}. \tag{4.13}$$

Glabowski's model [45] has extended the Wilkinson's formula to match the multi-rate traffic by

$$V_{i,j} = C_j - \sum_{l=1, l \neq i}^{C_j} c_{l,j}, \qquad j = 0, 1, 2, 3,$$

$$z'_{i,j} = 1 - a'_{i,j} + \frac{a_{i,j}}{V_{i,j}/m_{i,j} + 1 - a_{i,j} + a'_{i,j}}, \tag{4.14}$$

where $V_{i,j}$ is defined as the part of class j's capacity that is not occupied by the carried requests and $c_{l,j}$ is the lth flow's traffic carried by class j.

The same method used from (4.3) to (4.11) can be used in each individual class to compute the carried traffic and blocking probability of that class. The total blocking probability of the system is actually the dropping probability of the last class. Therefore, the system blocking probability β is equal to α_3. System utilization is computed by using the ratio of the total traffic carried by the system and the total system capacity, i.e.,

$$u = \frac{\sum_{j=0}^{3} c_j}{C_0 + C_1 + C_2 + C_3}. \tag{4.15}$$

4.4 Performance Evaluation

In this section, we present the analytical and simulation results of the *subframe-fit* and *isozone-fit* algorithms and validate their models presented in Sect. 4.3.

(1) Simulation Scenarios

The simulation tool is NS-2 version 2.33 [70] with added WiMedia DRP MAC module. An MPEG-4 video traffic generator [68] contributed to NS-2 is used to generate video streams at different data rates for homogeneous or heterogeneous traffic. To regulate the traffic according to the TSpec, each video traffic generator is attached to a twin-token bucket filter (i.e., a traffic shaper) [135]. The filter follows the traffic parameters of the video source, and we gather these parameters from the video trace file. For the simulation, a WPAN scenario is used for IPTV in-home

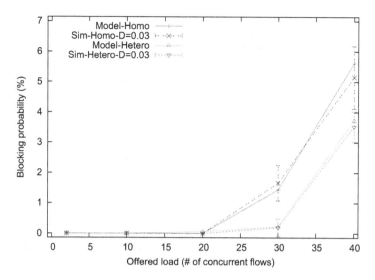

Fig. 4.6 Blocking probability of the subframe-fit algorithm and model [31]

distribution. Ten UWB stations are located on the circumference of a circle with a radius of 15 m. An access point (AP) is located at the center of the circle. All stations are in the one-hop transmission range of the AP, and we assume that the channel is ideal due to the short distance inside a house and the strong error recovery ability in UWB. All stations transmit video traffic to the AP at 480 Mbps.

(2) Subframe-Fit Performance

The analytical results are obtained following the approaches in Sect. 4.3 and are compared with the simulation results to validate the system model. In the analysis, requests are assumed to have a fixed size, and only the SI is considered. In simulation, both the queuing delay and the SI delay are considered for the maximum delay bound. The request size is determined by the traffic generator and the reservation algorithms. All simulation results presented in this section are the average of 15 runs. Error bars show 95 % confidence interval for the mean value. If the error bars are too small, they are not shown in the figures.

Figure 4.6 shows the blocking probability for the subframe-fit algorithm (Sim) and the subframe-fit model (Model) for both homogeneous (Homo) and heterogeneous (Hetero) traffic scenarios with a fixed delay of 30 ms. In the analysis, the subframe's length is computed based on the delay requirements of the flows, which limits the subframe-fit model to work only with the fixed delay value for all flows.

Based on the traffic specification, in the homogeneous scenario, the video flow requires on average six time slots per superframe for its bandwidth requirement. In the heterogeneous case we have used nine video flows, where stations pick one of them for transmission based on a uniform distribution. The flows require [2, 10] time slots. As shown in the figure, the subframe-fit model closely captures

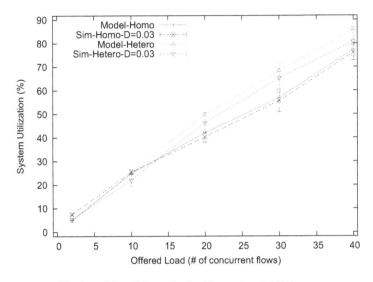

Fig. 4.7 System utilization of the subframe-fit algorithm and model [31]

the blocking probability of the algorithm. When the offered load is increased to 40 requests, if all requests were accepted, the system would be fully utilized. However, due to the delay bound constraints, not all requests can be accommodated, leading to a considerable blocking probability. The algorithm can perform better when there are variable-size flows, since they can "fill" the gap due to fragmentation, which explains that homogeneous scenarios indeed have worse performance.

Figure 4.7 shows both the analytical and simulation results of the system utilization. As discussed above, the subframe-fit model captures the properties of the real behavior. In the heterogeneous case, flows on average need six time slots but the superframe can be better utilized when the requests have variable sizes.

(3) Isozone-Fit Algorithm

In the following we present the results of both simulation and analysis for the isozone-fit algorithm, under both homogeneous and heterogeneous traffic scenarios as well as fixed and variable delay bounds. In the variable delay case, flows may have a delay requirement uniformly between [10, 50] ms.

Figures 4.8 and 4.9 show the blocking probability and system utilization of both homogeneous (Homo) and heterogeneous (Hetero) traffic with fixed and variable delay bound (D).

We can see from these two figures that the analytical model can capture the properties of both homogeneous and heterogeneous cases closely. Also the heterogeneous case has a better performance compared with the homogeneous scenario. Variable request sizes can better utilize the superframe. On the other hand, the performance with a variable delay bound is not as good as that with a fixed delay bound. Even though the superframe has empty time slots, flows with tight

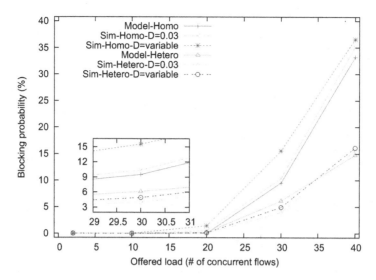

Fig. 4.8 Blocking probability of the isozone-fit algorithm and model [31]

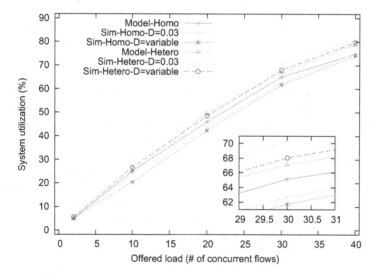

Fig. 4.9 System utilization of the isozone-fit algorithm and model [31]

delay bounds cannot be carried, and therefore we see a higher blocking probability for these cases (later we will show the improvement of isozone-fit to reduce the blocking probability).

In terms of the system utilization, the performance of the isozone-fit algorithm is not as good as the subframe-fit algorithm. The subframe-fit algorithm only considers the request sizes and their delay requirements whereas the isozone-fit specifically

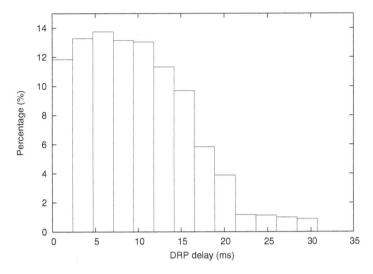

Fig. 4.10 Delay distribution of the isozone-fit algorithm [31]

follows the WiMedia MAC policies on block sizes and locations, which indicates that subframe-fit gives an optimistic bound for all WiMedia-compliant schemes.

(4) Delay Analysis

We study the delay that a flow will experience due to the reservation, which is an important performance metric in QoS provisioning. In the TSpec, flows specify the maximum tolerable delay. This delay is from the time when a packet arrives at the MAC layer till it is transmitted, which includes any queuing delay and the SI in DRP. Figure 4.10 shows the delay distribution in isozone-fit ($m = 6$ and $D = 30$ ms). As we can see from the figure, the majority of the packets encounters a delay due to DRP (both queuing delay and SI) well below the bound. Only 0.26 % of the packets (13 out of 5035 packets from the simulation trace) are slightly above the bound (the maximum packet delay is 30.7 ms), which is due to the EBW estimation in Sect. 4.3, but the delay outage probability is still below the chosen P_{loss} (1 %).

4.5   Reservation Performance Improvement

In [31], the authors proposed new algorithms to support video streaming over UWB networks. By introducing *cross-isozone allocation* and *on-demand compaction*, the performance of isozone-fit is further improved in terms of both blocking probability and system utilization. These methods help to accommodate more flows and also achieve a better utilization of the superframe.

4.5.1 New Reservation Mechanisms

(1) Cross-Isozone Allocation

The isozone-fit algorithm proposed in the previous section only makes reservations in one isozone of the superframe, following the regulation of WiMedia MAC standard. The algorithm can be extended by making reservations in multiple isozones. If a flow cannot be allocated in one isozone, instead of dropping the entire request to the next isozone with a higher isozone index, the request can be split into subrequests for the isozones of lower indexes, including the current isozone. By cross-isozone allocation, delay requirement can be met while achieving a higher system utilization.

To describe the idea, let us consider a flow request of size m with delay requirement of d. Due to the symmetric property of the isozone-fit reservation, reservation blocks of a request are evenly distributed in the zones of each isozone. Therefore a request can be considered as

$$m = x_0 + 2x_1 + 4x_2 + 8x_3, \tag{4.16}$$

where x_j is the block size of the request in isozone j and isozone j has 2^j columns (or zones). Based on the value of d and the natural SI of each isozone, the block size of x_j, where j is the isozone index with the SI less than or equal to the requested delay, should be nonzero. Block sizes with indexes higher than j are initially set to zero. To implement this heuristic approach, a recursive search method is added to the algorithm, which is designed according to the request size and isozone index. The pseudo-code of this method is listed in Algorithm 1.

Algorithm 1 Cross-isozone fit (m_i, j)

Require:
    m_i: request size of flow i; j: starting isozone index
    c_j: number of the columns of isozone j
    $x_{j,k}$: allowable available time slots in column k of isozone j
Ensure: cross-isozone allocation
    while request not carried and $0 \leq j \leq 3$ **do**
        if $c_j \times x_{j,k} \neq 0$ **then**
            $y = m_i - c_j \times x_{j,k}$
            if $y > 0$ **then**
                "Cross-isozone fit $(y, j - 1)$" // use previous isozones
            else
                "Done"
            end if
        else
            $j = j + 1$ // drop to the next isozone
        end if
    end while

As shown in the algorithm, a flow request is forwarded to isozone j based on its delay requirement. Since the allocation method is symmetric across the columns of each isozone, the time slot availability is also symmetric for the columns of each isozone. If the time slots in isozone j are available, they are used to make the reservation. If part of the request cannot be carried, the time slots from the previous isozones are used. This search method is performed by recursively calling the same procedure with a smaller isozone index. If the available time slots in isozone j and isozones with lower indexes are not enough, the entire flow is dropped to the next isozone with a higher index. The search method stops when either a request is carried or there are no more isozones to search through. Since the method is applied on each flow upon its arrival, the solution is not necessarily the overall optimal solution with full knowledge.

(2) On-Demand Compaction

So far in all the proposed reservation algorithms, flows use the same time slots for transmission during their existence in the system. The authors in [31] have proposed the reallocation strategy to further improve the performance of the algorithms. The same technique is used in the memory management of operating systems and it is called *compaction*. It is the technique of relocating all occupied areas of memory to one end of the memory so as to get one large block of free memory space. The same idea can be used here but the delay requirement of the flows should still be satisfied after reallocation. This means if all or part of a request has reserved time slots in a certain isozone, after compaction, it should still be in the same isozone so the delay constraint is not compromised. Therefore, in this process all of the allocated blocks in each column of an isozone are moved to one "end" of the column. Consequently, all of the free segments are combined into a single large free block extending to the other end of column. This helps to better utilize the superframe.

Superframe can be compacted under the following conditions: (1) as soon as a flow stops transmission and leaves the piconet, (2) when a new request cannot be allocated in the superframe, and (3) at fixed time intervals. Most of memory compaction in operating systems is performed "on-demand". In [31], the authors have used the same technique which is believed to be the most effective way of compaction. Due to the symmetric structure of isozone-fit, the compaction method can be applied uniquely in a distributed manner at each station.

4.5.2   Performance Evaluation

In this section, we show the simulation results of the improvement methods discussed in the previous subsection. The results of the heterogeneous case with fixed delay are presented, which is a good representative of the other cases.

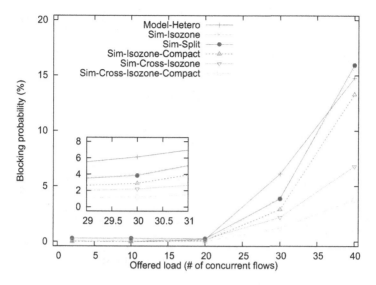

Fig. 4.11   Blocking probability of the isozone-fit improvement [31]

(1) Blocking Probability

Figure 4.11 shows the blocking probability of both simulation (Isozone, Cross-isozone, and Compaction) and analysis (Model) of the heterogeneous scenario with fixed delay bound of 30 ms. We compare the results of the isozone-fit and the cross-isozone algorithm, and also apply the on-demand compaction strategy to both of them. It can be observed that the compaction improves the results of both the isozone-fit and cross-isozone algorithms. The best performance is achieved when we have the cross-isozone fit with the on-demand compaction, which is very close to the bound given by subframe-fit.

(2) System Utilization

The same improvement can be seen for system utilization in Fig. 4.12. Compared with the original isozone-fit algorithm, the cross-isozone algorithm with on-demand compaction improves the system utilization by more than 10 %.

(3) Running Time

Complexity is an important aspect of all these reservation algorithms and improvement methods. Although heuristic in nature, the proposed algorithms do not give reservations when TSpec cannot be met due to the way of their construction. Also they always terminate within a finite amount of time due to the limited allocation zones. The authors in [31] claimed that subframe-fit is computationally simple because it does not fully follow the WiMedia standards, whereas isozone-fit closely captures the WiMedia MAC. The running time of each reservation algorithm is evaluated by simulation as following.

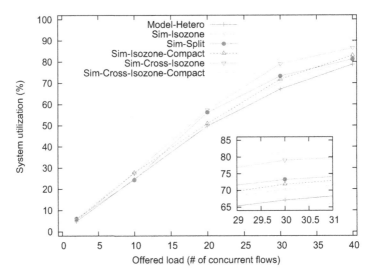

Fig. 4.12 System utilization of the isozone-fit improvement [31]

Table 4.1 Average reservation processing time per flow

Algorithm	Subframe-fit	Isozone-fit	Cross-isozone-fit
Processing time (μs)	1.18	2.89	2.39
Algorithm	Isozone + Compaction	Cross-isozone-fit + Compaction	
Processing time (μs)	3.14	2.60	

Table 4.1 shows the average reservation processing time per flow for heterogeneous traffic with a fixed delay of 30 ms. The simulations were performed on a quad-processor Intel(R) Xeon(R) computer with 2.66 GHz CPU, 4 GB RAM, and Scientific Linux SL release 5.2. Subframe-fit has the smallest running time as expected. Compared with isozone-fit with compaction-only, all other isozone-fit algorithms reduce the running time due to the well-structured superframe which makes the reservations easier to fit.

4.6 Summary

In this chapter, we have discussed the principles and challenges in the reservation-based MAC for QoS provisioning. With emphasis on the distributed reservation protocols, we use WiMedia UWB MAC as an example and analyze two algorithms, *subframe-fit* and *isozone-fit*. Equivalent bandwidth theory, non-Poissonian traffic decomposition, and the approximation of overflow traffic in multi-class-traffic, multi-class-server systems, have been used for analytical modeling of the channel time reservation. Network simulation and MPEG-4 traffic generator have been performed to validate the models. Although subframe-fit is not fully WiMedia-compliant, it gives us an optimistic performance bound, which we can use as a reference to improve the performance of isozone-fit. This has been achieved by introducing cross-isozone allocation and on-demand compaction to isozone-fit, and the performance improvement has been confirmed by the simulations in Sect. 4.5.2.

In this chapter, the EBW approach is utilized to compute the number of time slots required for reservation. EBW is computed from the flow's TSpec (flow's mean and peak rate). However, flows do not always transmit at their peak rate and will have more time slots reserved than they actually need for data transmission. One solution is to divide each video stream into two parts and try to reserve time slots at the average data rate of a video stream (e.g., for the important packets in the base coding layer) while other packets are carried through the contention-based channel access. In this way, the critical part of video streams can be carried through contention-free transmission opportunities to guarantee QoS and meanwhile a better channel utilization may be achieved. This "hybrid" approach will be discussed in details in the next chapter.

Chapter 5
Hybrid Medium Access for Multimedia Services

To reserve or not for bursty video traffic over wireless access networks has been a long-debated issue. Reservation can ensure QoS provisioning at the cost of lower resource utilization. Contention-based MAC is more flexible and efficient in sharing resources for bursty traffic by a higher multiplexing gain, but the QoS may degrade severely with the increase of traffic load. In wireless networks using hybrid MAC, nodes can reserve time periods inside scheduling cycles and the time not reserved can be used by all stations through contention-based access. The hybrid MAC is attractive because each video source reserves well below its peak data rate and uses contention-based media access to transmit the remainder/bursts of the traffic. Thus, satisfactory QoS may be provided by resource reservation and high channel utilization can be achieved due to the multiplexing gain in the contention periods. In this chapter, we first introduce the hybrid hard and soft-reservation schemes and then present the analytical models. We illustrate how to use the mean value analysis approach to calculate the collision probability and the average service time of a frame. Furthermore, using the standard WiMedia MAC protocols as an example, extensive simulations using NS-2 and real video traces are given which verify the analysis and demonstrate the effectiveness of the hybrid MAC. The results show that the hybrid MAC, especially with soft reservation, has much better delay performance and higher capacity (supporting more video streaming flows) than the contention or reservation-only MAC.

© The Author(s) 2017   103
R. Zhang et al., *Resource Management for Multimedia Services in High Data Rate
Wireless Networks*, SpringerBriefs in Electrical and Computer Engineering,
DOI 10.1007/978-1-4939-6719-3_5

5.1   Hybrid Approach

5.1.1   Hard-Reservation Dual-Buffer Hybrid Medium Access

(1) Issues in Contention or Reservation-only MAC

To share wireless resources by video traffic, the contention and reservation-based channel access approaches have their pros and cons. The former is flexible and efficient in sharing resources for bursty traffic and can achieve a certain level of multiplexing gain. However, it suffers from an intrinsic weakness: when there are multiple sources in the system, even though the sum of their average data rates is below the channel capacity, excessive queueing delay and frame losses are possible due to collisions and backoff procedures, as discussed in Chap. 3. This may lead to a considerable degradation of video quality. The latter (i.e., contention-free) is often preferred for video streaming, which guarantees the channel access opportunity to satisfy the delay requirement, as mentioned in Chap. 4. However, due to the burstiness of video traffic, if reservations are made at the peak data rate to minimize queuing delay, over-reservation leads to significant waste of bandwidth; if made at the average data rate, considerable queuing delay and frame losses can occur during traffic bursts. Therefore, neither reservation nor contention alone is an effective approach for supporting video streaming in wireless networks.

(2) Hybrid Contention/Reservation Channel Access Approach

Newer wireless standards adopting a hybrid approach have emerged, which allow both reservation and contention-based access, for instance, the MAC protocols defined in the ECMA-368 [69] and IEEE 802.15.3 [86] standards for high-rate WPANs.

We study the general contention/reservation interleaved hybrid MAC approaches in this chapter. Each access *scheduling cycle*, denoted by T_Y, is composed of multiple contention and reservation periods which may be interleaved, as shown in Fig. 5.1. A station can send traffic in its reserved periods and compete for TXOPs during the contention periods following the contention rules such as CSMA/CA.

The stations in the network can reserve channel access time with the coordination of a central controller (e.g., following the IEEE 802.15.3 standard) or by themselves in a distributed manner (e.g., following the WiMedia ECMA-368 standard). Considering fairness, suppose that each station reserves a duration of T_S in a scheduling cycle. The time interval between two consecutive reserved periods, denoted by T_C, is available for contention-based channel access. It is desired to reserve time slots uniformly allocated in the scheduling cycles for one flow to reduce delay variation and queue length. Thus, each multimedia flow can obtain guaranteed channel access periodically and also compete during the contention periods.

During the reserved period, the reservation owner can transmit backlogged frames back-to-back and then receive a block acknowledgment (B-ACK) at the end of the burst. The data frames are separated by an minimum interframe spacing

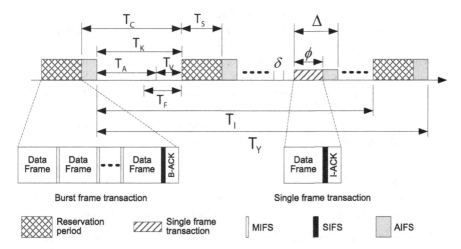

Fig. 5.1 Hybrid channel access approach with hard-reservation [157]

(MIFS), while the B-ACK frame is separated by an SIFS to allow the transmission/reception mode switch of the involved stations. Following the B-ACK is a guard time (GT) in order to separate the reservation and contention periods. The burst size is determined such that the burst transaction (including B-ACK and GT) should be finished within the reserved time. In the burst mode, the duration for a frame transaction is $\phi_B = T_{DATA} + MIFS$. Such burst transmission can reduce protocol overheads and increase channel utilization.

During T_C, a station obtains the TXOP when the backoff counter is decreased to zero. A single transaction with an immediate-acknowledgement (I-ACK) may be completed in the TXOP. The duration is $\phi = T_{DATA} + SIFS + T_{ACK}$, where T_{DATA} and T_{ACK} are the transmission time of a data frame and the I-ACK frame, respectively.

The novelty of hybrid MAC is to reserve well below the peak data rate and handle traffic bursts using contention-based channel access, aiming to fully utilize the reserved channel time and reduce the traffic load for competition during the contention periods (reducing the collision probability and backoff time). Consequently, the hybrid approach can provide better QoS and also increase network capacity (i.e., supporting more flows) by more efficient channel contention.

(3) Conflict Avoidance

The contention-based access in hybrid MAC follows the CSMA/CA and exponential backoff schemes such as the IEEE 802.11 DCF protocol. However, the contention is even more complicated compared with the traditional contention-only MAC due to the presence of the reserved channel time slots.

1. A station shall freeze its backoff counter once the medium becomes busy (during frame transactions) or unavailable for contention (during reserved periods). The station shall wait the channel to be in the contention-access period and sense

idle for an *AIFS* before starting to decrement the backoff counter. Therefore, the reservations enlarge the backoff slot and service time of the contention-based access considerably.

2. When a station obtains a TXOP, it needs to ensure that the whole frame transaction finishes at least one *SIFS* plus one *GT* before the beginning of the incoming reserved period. Otherwise, the TXOP has a *conflict* with the reservation. The time interval of $T_F = \phi + SIFS + GT$ is thus called *conflict time*. If a station obtains a TXOP during T_F, it should avoid conflict with the incoming reserved time slot, following one of the two conflict avoidance strategies:

   • *Hold-on strategy*: the station just holds on and transmits immediately after the reserved time plus *AIFS*.
   • *Backoff strategy*: the station invokes the next backoff stage by selecting a new backoff counter, i.e., having a virtual collision with the reservation.

It has been shown that the backoff strategy results in less frame service time and higher throughput with a moderate traffic load [112]. Suppose that several stations obtain TXOPs during T_F. According to the hold-on strategy, their transmissions right after the reserved period will collide. On the contrary, if the backoff strategy is adopted, the stations will initiate another round of backoff, so their collision probability is reduced. The performance of the two conflict avoidance strategies will be compared by simulations in Sect. 5.3.

As shown in Fig. 5.1, the stations can perform backoff inside the time interval between the end of the previous reservation period plus *AIFS* and the beginning of the next reservation. This interval is denoted as T_K, i.e., $T_C - AIFS$. T_K is further divided into T_A and T_V. The frame transactions initiated in the current contention period are completed during T_A (called *access time*). Because an ongoing transaction may finish inside T_F before the next reservation period, the time period left in T_C, i.e., T_V (called *vulnerable time*), is actually smaller than T_F.

(4) Dual-Buffer Architecture

The hybrid MAC approach can be accompanied with two different buffering architectures for video streaming. The single-buffer and the dual-buffer architectures are illustrated in Fig. 5.2a and b, respectively. Using the former architecture, when the reserved period becomes available for a flow, frames accumulated in the buffer are transmitted up to the maximum burst size. Otherwise, the remaining frames compete with other flows for channel access during the contention periods.

One major issue with the single-buffer architecture is as follows. Due to the asynchronous nature between video traffic and reservation schedule, it is possible that there are not enough frames to send in the buffer during a reserved period and thus the reserved time is under-utilized. Consequently, the load and the contention level during the contention periods may be still quite high.

Using a dual-buffer architecture can solve this problem. As shown in Fig. 5.2b, the size of the R-Buffer equals the maximum number of frames that can be transmitted in a reservation slot. The arrived frames are first backlogged in the R-Buffer to wait for the next reserved period. When it is full, newly arrived

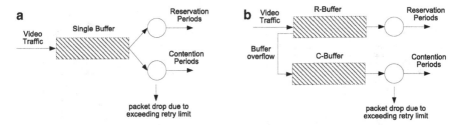

Fig. 5.2 Buffering architectures for the hybrid MAC. (**a**) Single buffer architecture (**b**) Dual buffer architecture

frames are stored in the C-Buffer which will contend for the channel access during contention periods.[1] The main advantage of the two-buffer architecture is that the number of frames in the reserved periods is maximized so as to maximize the utilization of the reserved periods. Consequently, the frame arrival rate and competition level during contention periods are reduced and thus the collision probability and service time are reduced compared with the contention-only MAC.

However, the dual-buffer architecture introduces two issues:

1. It causes frame re-ordering at the receiver side. In one case, the follow-on frames may have been delivered but the earlier frames are still in the R-Buffer waiting for the reserved time. However, the scheduling can be set such that the reservation interval is smaller than the delay jitter bound. In the second case, it is also possible that the earlier frames are still waiting in the C-Buffer due to the random backoff procedure and follow-on frames have been received through reserved period. For this case, we can restrict the competition in the contention period by using the admission region (as will be shown in Sect. 5.4) to guarantee that the frame delay in the C-buffer will also not exceed the delay jitter bound. Thus, the reordering due to the two-buffer architecture does not affect the video streaming performance.
2. Part of the reserved time is still wasted due to the traffic burstiness. To solve this problem, we can utilize soft-reservation by which the unused reserved time is released and the other stations can compete to utilize the released time. The soft-reservation approach will be described in the next subsection.

We need to quantify and optimize the hybrid MAC performance carefully to ensure the QoS for video flows. However, the existing analytical models for traditional contention-only MAC cannot be directly applied to hybrid MAC due to the difference presented above, whose model and analysis will be presented in Sect. 5.2.

---

[1]Note that if transmission errors happen in the burst transaction during reserved periods, the frames that need to be retransmitted can be queued in the C-Buffer. In this chapter, we only consider collisions and ignore transmission errors to simplify the analysis.

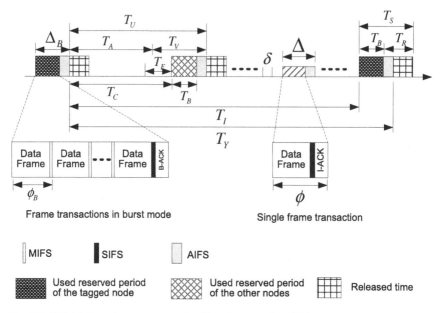

Fig. 5.3 Hybrid channel access approach with soft-reservation [156]

5.1.2 Soft-Reservation Hybrid Medium Access

Due to the burstiness of video traffic, it is possible that the reservation owner has no more frames to transmit during its reserved time, resulting in the waste of channel time. In this subsection, we consider *soft reservation*, where the owner implicitly releases the unused reserved time for other stations to access by contention. The model of soft-reservation hybrid MAC is shown in Fig. 5.3. The reservation owner may transmit the backlogged frames in a burst transaction from the beginning of its reserved periods (as presented in the previous subsection, the transmissions from the other stations are suppressed before the reserved periods due to the conflict avoidance strategy). Similarly to the hard-reservation approach, frame transactions in the burst mode during reserved periods finish with a B-ACK, while a single-frame transaction during contention periods finishes with an I-ACK. The durations of a frame transaction in the two cases are ϕ_B and ϕ, as mentioned earlier.

Other stations keep sensing the channel. Because they have to wait for the channel to be idle for AIFS (which is larger than MIFS and SIFS), they cannot perform backoff or transmit during the burst transaction. If the reservation owner has no more frames to transmit, the channel will be idle for an AIFS after the B-ACK and then the other stations start or resume backoff procedures, the same as that during the contention periods. Thus the unused reservation is released automatically. Furthermore, the reservation owner can have a higher priority to regain the channel and transmit new frames, by just waiting for a shorter IFS during its released time. Therefore, the soft reservation is easy to implement.

This soft-reservation approach leads to more efficient channel utilization and better network performance (e.g., smaller frame service time) than the hard reservation (which does not release the unused time). In addition, the dual buffer architecture is unnecessary for soft reservation because the unused reservation period can be released.

5.2 Mean Value Analysis

In this section we study the analytical models for hybrid MAC considering the interactions of reservation and contention periods. The reservation-based access is deterministic, while the behavior of the contention-based channel access is significantly affected by reservation, which is essentially different from the traditional contention-only MAC (such as the IEEE 802.11 DCF). Therefore, we focus on the performance analysis of the contention-based access.

Using the *mean value analysis* method, we evaluate the average values of the system variables, such as station transmission probability, collision probability, and frame service time, without considering the details of the stochastic backoff process. We first present the model for the hard-reservation hybrid MAC with saturated traffic case, and then show how to extend it to derive the tight performance bounds for the unsaturated case. Secondly, the analytical model for the soft-reservation scheme is developed. The approaches are of low computational complexity so they can be used for on-line admission control and optimizing the per-flow reservation.

5.2.1 Hard-Reservation Hybrid MAC with Saturated Stations

(1) Network Model

We consider a collision domain with N active stations and an ideal channel condition without transmission errors or capture effect. Stations reserve channel time with the coordination of a central controller (e.g., the IEEE 802.15.3 standard) or by themselves in a distributed manner (e.g., the WiMedia ECMA-368). Considering fairness, all stations reserve sequentially and each reserves a duration of T_S in a scheduling cycle, as shown in Fig. 5.1. The time interval between two consecutive reserved periods, denoted by T_C, is available for contention-based channel access. All MAC frames are assumed to have the same length.

(2) Transmission Probability

Define the time for a sensing station to decrement its backoff counter as one generic time slot, which may have a various duration according to different channel states as discussed later. The key performance parameters are the collision probability and the service time of a frame, where the service time includes both the backoff and transmission time slots.

Let $E[R]$ and $E[B]$ denote the average number of transmission slots and backoff slots experienced by a frame, respectively. Given that a station is busy, i.e., the station is performing backoff or transmitting, the probability to transmit in a slot equals

$$\tau = \frac{E[R]}{E[R] + E[B]}. \tag{5.1}$$

Let p denote the collision probability for the tagged station when it sends a frame. After collision, the station tries to retransmit the frame. The number of transmission trials of a frame follows a truncated geometric distribution with the success probability of $1 - p$. Suppose that the retry limit is K (the indexes of the backoff stages are $k = 0, 1, 2, \cdots, K$) and the CW in the kth stage is $[0, W_k]$. Similar to the derivations in Sect. 3.3.1, $E[R]$ and $E[B]$ can be obtained by

$$\begin{cases} E[R] = \sum_{k=0}^{K} p^k, \\ E[B] = \sum_{k=0}^{K} \left(p^k \frac{W_k}{2} \right), \end{cases} \tag{5.2}$$

where W_k and $\frac{W_k}{2}$ are the contention window size and the average number of backoff slots in the k-th backoff stage, respectively.

(3) Contention Access Period

During T_A, a time slot can have three states.

First, a slot may be idle if no station transmits. The duration of an idle slot is δ and, within T_A, the probability of a slot to be idle is a_A.

Second, a slot may contain a frame transaction, either successful or with collision, with probability of b_A. The duration of this type of slots is $\Delta = \phi + AIFS$.

Third, if a frame transaction begins during the interval of $[T_F, \Delta]$ before a reservation period (as defined in ECMA-368, $\Delta > T_F$), then after the frame transaction, the reserved time period will begin within an AIFS. Thus, all the contention stations have no chance to perform backoff after this frame transaction. We can regard the reservation period together with the frame transaction as one time slot with probability of b_{AD} and the duration of Δ'. We approximate that, if such a scenario happens, the beginning time of the frame transaction is uniformly distributed inside $[T_F, \Delta]$ before the reservation. Thus, the average duration of this type of transaction slots is $\Delta' = \frac{\Delta + T_F}{2} + T_S$. Further, given that a frame transaction occurs within T_A, the probability for the transaction begins during $[T_F, \Delta]$ is $\frac{\Delta - T_F}{T_A}$. In this scenario, we also have $T_V = 0$.

The probabilities for a slot to be in the three states are given by

$$\begin{cases} a_A = (1 - \tau)^N, \\ b_{AD} = (1 - a_A)\frac{\Delta - T_F}{T_A}, \\ b_A = 1 - a_A - b_{AD}. \end{cases} \tag{5.3}$$

Thus, during T_A, the average slot duration is

$$S_A = a_A \delta + b_A \Delta + b_{AD} \Delta'. \tag{5.4}$$

If a transaction begins inside the time interval $[T_F, T_F + \Delta]$ before the beginning of a reservation period, the transaction will end inside the conflict time, which makes the vulnerable time T_V smaller than T_F on average. If a transaction begins inside the interval of $[T_F, T_F + \Delta]$, we approximate that its starting time is uniformly distributed, so the average vulnerable time is $\frac{T_F}{2}$. However, if no transmission occurs inside the interval, the vulnerable time is T_F. The number of idle slots in the interval $[T_F, T_F + \Delta]$ is $\Gamma_\Delta = \frac{\Delta}{\delta}$. The probability of no transmission initiated during the interval is $a_A^{\Gamma_\Delta}$. The average duration of the vulnerable time is[2]

$$T_V = \left(1 - a_A^{\Gamma_\Delta}\right) \frac{T_F}{2} + a_A^{\Gamma_\Delta} T_F = \left(1 + a_A^{\Gamma_\Delta}\right) \frac{T_F}{2}. \tag{5.5}$$

Then, the average duration of T_A can be obtained by $T_A = T_K - T_V$, and the average number of slots in T_A is $\Gamma_A = \frac{T_A}{S_A}$.

(4) Vulnerable Time and Reservation Period

Because the contention stations cannot transmit during T_V and the following reserved period, we consider these two periods together. A slot inside T_V has two states. The upcoming reservation period arrives during the last idle slot, which is thereby enlarged by T_S plus AIFS. We assume that the arrival time of the following reserved period is uniformly distributed inside the last idle slot in T_V. Thus, the average duration of the enlarged slot is

$$\delta_D = \frac{\delta}{2} + T_S + AIFS. \tag{5.6}$$

All the other slots within T_V are idle slots with the duration of δ. The average duration of T_V is given in (5.5) and hence the average number of slots can be estimated by $\Gamma_V = \frac{T_V}{\delta}$. Finally, given a slot within the vulnerable time, the probability to have the duration of δ (i.e., not the last one) is $g = \frac{\Gamma_V - 1}{\Gamma_V}$.

(5) Generic Channel Slot

Considering the total number of generic slots during T_K, the probability for a generic slot to be in the vulnerable time is

$$h = \frac{\Gamma_V}{\Gamma_A + \Gamma_V}. \tag{5.7}$$

---

[2]For easy presentation, we reuse the notations of T_A, T_S, etc., so they also represent the averages of the corresponding durations in the equations in this section.

In summary, from the viewpoint of the entire channel time, the generic channel slots have four states: idle slots within T_A and T_V periods and with the duration of δ and the probability of a; frame transaction slots within T_A and with the duration of Δ and the probability of b; after the frame transaction, if the incoming reservation period starts within an AIFS, the transaction slot is combined with the reservation period and they are regarded as one slot, with the duration of $\Delta_D = \Delta' + AIFS$ and the probability of b_D; and the last idle slot inside T_V is combined with the following reservation period as one slot, with the duration of δ_D and the probability of a_D. The state probabilities are given by

$$\begin{cases} a &= hg + (1-h)a_A, \\ b &= (1-h)b_A, \\ b_D &= (1-h)b_{AD}, \\ a_D &= h(1-g). \end{cases} \tag{5.8}$$

Thus, the average duration of a generic slot is

$$S = a\delta + a_D\delta_D + b\Delta + b_D\Delta_D. \tag{5.9}$$

(6) Collision Probability

Given that the tagged contention station transmits in a slot outside the conflict time, collision will happen if any other stations also transmit in the same slot. For saturated stations, the probability to transmit in a slot is τ, given in (5.1). On the other hand, if the TXOP is obtained within the vulnerable time, the station has to defer the transmission according to the conflict avoidance strategy.

Using the backoff strategy, the conflict can be regarded as a *virtual collision* with the incoming reserved time period, which results in a new backoff stage. Because the probability for a slot being in the vulnerable time is h given in (5.7), the overall collision probability (including virtual collisions) is

$$p = 1 - (1-h)(1-\tau)^{N-1}. \tag{5.10}$$

By solving Eqs. (5.1)–(5.10) numerically, we can obtain the mean values of the system variables, such as p, τ, etc.

For the hold-on strategy, a virtual collision will not result in a new backoff stage, but the station will experience collisions when more than one stations obtain TXOPs within the vulnerable time. In this case, the collision probability is

$$p = 1 - (1-h)(1-\tau)^{N-1} - h(1-\tau)^{(N-1)\Gamma_V}. \tag{5.11}$$

Similarly, we can numerically obtain p and τ for the hold-on strategy. We will compare the performance of the two conflict avoidance strategies in Sect. 5.3.

(7) Average Frame Service Time and Station Throughput

Because the transmission of a frame experiences $E[B] + E[R]$ generic slots, the average service time is

$$\zeta = (E[B] + E[R])S. \tag{5.12}$$

Since only one frame is (re-)transmitted during the frame service time on average, the rate to send a new frame is $\frac{1}{\zeta}$. When all the transmissions (up to the specified retry limit, K) of a frame have failed, the frame will be discarded. Therefore, the per-station throughput (bps) can be obtained as

$$\eta = \frac{L_P}{\zeta} \left(1 - p^K\right), \tag{5.13}$$

where L_P is the payload size of a frame in bits.

5.2.2 Hard-Reservation Hybrid MAC with Unsaturated Stations

Different from the saturated-station case, an unsaturated station only contends for channel access when its buffer is non-empty. The key parameter is the probability that a station is busy in a selected generic slot, called *busy probability*.

(1) Correlated Channel Access

The most challenging issue for the analysis with unsaturated stations comes from the fact that the assumption for the unsaturated stations to be busy *independently* is not valid. This is because the channel sensing effort from a station may be blocked by the ongoing transaction or the reservation period. As a result, the probability for another station to be busy given that the tagged station is busy (i.e., the conditional busy probability) is higher than the probability for another station to be busy in a randomly selected slot (i.e., unconditional busy probability).

In this section, we extend the mean value analysis to the hybrid MAC with unsaturated stations. Because it is very complicated, if not impossible, to obtain the accurate conditional busy probability of another station given that the tagged station is busy, we develop the lower and upper bounds of the system performance. As shown by the numerical results, these two bounds are tight and they *converge* when the stations become saturated.

(2) Lower Bound

Because the probability for another station to be busy conditioned on that the tagged station is busy is larger than the unconditional busy probability, we can obtain the lower bound of the collision probability by using the unconditional busy probability in a randomly selected slot. The symbols with the superscript ′ represent the lower-bound parameters.

First, the unconditional busy probability for a station in a randomly selected slot is

$$\rho' = \min \left\{ \frac{(R' + B')S'}{\mu}, \, 1 \right\}, \tag{5.14}$$

where μ is the average arrival interval of frames and S' is the average duration of a generic slot for the entire channel time. In (5.14), $\frac{\mu}{S'}$ gives the average number of generic slots between two consecutive arrivals. Given the station is busy, the transmission probability is τ', the same as the saturated case in (5.1). Thus, an unsaturated station transmits in a generic slot inside the contention period with the probability $\rho'\tau'$. The probabilities of the three states of a slot inside T_A are changed to be

$$\begin{cases} a'_A = (1 - \rho'\tau')^N, \\ b'_{AD} = (1 - a'_A)\frac{\Delta - T_F}{T_A}, \\ b'_A = 1 - a'_A - b'_{AD}. \end{cases} \tag{5.15}$$

In addition, the collision probability is changed to be

$$p' = 1 - (1 - h')(1 - \rho'\tau')^{N-1}. \tag{5.16}$$

Following the procedure in Sect. 5.2.1, we can obtain the unsaturated versions of the other equations, which have the same form as those for the saturated case. Combined together with (5.14)–(5.16), the mean values of all the system variables can be solved. Finally the frame service time ζ' is obtained similarly to that in (5.12).

Different from the saturated case, the throughput for unsaturated stations depends on the incoming traffic load. Here we ignore the limit of MAC buffer size and frame drops are caused by exceeding the retry limit only. Since the lower bound of the collision probability is p', the upper bound of the station throughput is

$$\eta' = \frac{L_P}{\mu} \left[1 - (p')^K \right]. \tag{5.17}$$

(3) Upper Bound

In a network with all stations unsaturated, there must be some time periods that all stations are non-busy. In other words, the probability that there is no station busy should be larger than zero. Therefore, the upper bound of the collision probability is obtained by assuming that at any time moment, there is at least one station being busy. The symbols with the superscript $''$ represent the upper-bound parameters.

Under this hypothesis, the probabilities of the three states of a slot inside T_A becomes

$$\begin{cases} a''_A = (1 - \tau'')(1 - \rho''\tau'')^{N-1}, \\ b''_{AD} = (1 - a''_A)\frac{\Delta - T_F}{T_A}, \\ b''_A = 1 - a''_A - b''_{AD}. \end{cases} \tag{5.18}$$

Except the new version of (5.18), all the other equations have the same forms as those for the lower bound case presented in the previous subsection. For example, the station busy probability ρ'' and the collision probability p'' can be calculated similarly using (5.14) and (5.16), respectively. By solving this new equation set, we can obtain the mean values of the system variables.

Note that, the average slot duration conditioned on that a station is busy is larger than the average duration of a randomly selected slot. Therefore, S'' (obtained by assuming there is at least one station busy) is the upper bound of the average duration of a generic slot in a network with all stations unsaturated. Finally, the lower-bound of the station throughput can be obtained in a similar way.

(4) Asymptotic Property

An unsaturated station can be driven into saturation when its traffic load increases or when more channel time is reserved. When the portion of reserved time increases, the occurrence of conflict (virtual collision) and also the backoff slot duration will increase. Consequently, the frame service time increases. When the service time is larger than the frame arrival interval, the station becomes saturated.

When the stations in the network become saturated, the approximation for the lower-bound model (i.e., the busy probability of another station conditioned on that the tagged station is busy is equal to the unconditional busy probability) becomes accurate because both probabilities approach 1. On the other hand, with all stations becoming saturated, the hypothesis for the upper-bound model that there is always at least one node being busy, is satisfied. Therefore, the lower-bound and upper-bound both converge to the analytical results of the saturated case. Meanwhile, the convergence of the two bounds indicates that the contention stations become saturated.

5.2.3 Soft-Reservation Hybrid MAC with Unsaturated Stations

The analysis of the soft reservation with a single buffer is more complicated than the hard reservation with dual buffer. The durations and traffic load of the reservation and contention periods are random and interact with each other. For example, the service time of the contention-based access affects the burst sizes in the reserved periods, because the service rate during the contention periods determines the number of frames to transmit during reserved time. Meanwhile, the durations of the reserved periods used by the owners affect the backoff slot length and the service time of the contention-based access.

(1) Network Model

The network model is the same as that for the hard-reservation hybrid MAC. We consider a single-hop network with N stations (or N flows) and the frame arrival for each flow follows a Poisson process with the arrival rate of λ frames/s. As shown in Fig. 5.3, during T_C, a station obtains a TXOP when the backoff counter is decreased to zero, during which a single transaction with an I-ACK may be completed. The duration of a TXOP is $\phi = T_{DATA} + SIFS + T_{ACK}$, where T_{DATA} and T_{ACK} are the transmission time of a data frame and the I-ACK frame, respectively, as defined in Sect. 5.1.1. As for the hard-reservation model, the conflict time is $T_F = \phi + SIFS + GT$. Because a transaction which is initiated before T_F may finish inside T_F, the vulnerable time where the stations encounter conflict is smaller than T_F on average. The time interval before T_V is the contention-based access time, denoted by T_A in Fig. 5.3. Note that the owner station of the incoming reservation does not perform conflict avoidance and can transmit immediately.

As mentioned in Sect. 5.1.2, the reservation owner has a higher priority during the released time. For easy explanation, we incorporate the frames transmitted during the released time in the burst transaction. This simplification does not affect the accuracy of the analysis due to the following: if the network is lightly loaded, the probability to have new frames arrive during the released time is quite small; if the network is heavily loaded, the duration of the released time becomes quite small and the impact of this simplification can be ignored.

(2) Renewal Process

According to the system model in Fig. 5.3, the channel time is divided into two phases from the viewpoint of a tagged station: (1) the burst transactions inside reserved periods, and (2) the interval between the tagged station's two consecutive reserved periods where it accesses the channel by contention. The average durations of the two phases are denoted by T_B and T_I, and the average numbers of frames delivered are V_B and V_I, respectively. Thus, the average time released by the owner in a reserved period is $T_R = T_S - T_B$.

Furthermore, inside the scheduling cycle T_Y, the intervals of T_A, T_V, and the burst transactions (with the average duration of T_B plus the AIFS) occur alternatively and repeatedly. Thus, we combine them together as T_U (called *unit time*). T_U is the small cycle and there are N cycles of T_U inside an interval of T_Y.

The system can be regarded as a renewal process, and T_Y is the renewal cycle. Considering the stability of the network, we classify the traffic situation into three levels:

1. *Lightly loaded:* The frame arrival rate is so small that the average service time during the contention period (denoted by ζ) is smaller than the frame arrival interval (denoted by μ and $\mu = 1/\lambda$), i.e., $\zeta < \mu$. Thus, the reserved periods are mostly unused and released.
2. *Heavily loaded:* The frame arrival rate is larger and $\zeta > \mu$. Some frames are backlogged in the buffer during contention periods, but they are delivered during reserved periods. The network is still stable because all frames are delivered by the end of a scheduling cycle.
3. *Overloaded:* The frame arrival rate is so large that ζ becomes much larger than μ. There are so many frames backlogged in the buffer during a contention period that the number of backlogged frames plus those arriving during the reserved period is larger than the maximal burst size. Thus, the reserved periods are fully used and the queue length in the buffer will keep increasing and the network is unstable.

For the first and second levels, the network is stable and the frame arrival rate is guaranteed smaller than the overall frame service rate (including by both contention-based and reservation-based transmissions). We have the following necessary condition

$$T_Y \lambda < V_I + V_B. \tag{5.19}$$

Consequently, the queue length of a station should be 0 on average at the end of its own reserved period. In other words, *on average* the frames that accumulate in the buffer during T_I and arrive during T_S must be delivered by the end of the station's reserved period.

(3) Traffic Transmitted in Different Phases

To estimate V_I, let ζ denote the average service time of contention-based access during T_I (as defined earlier). If a station has frame(s) arriving during ζ before its own reserved period, the frame(s) will be delivered inside the incoming reserved period. Thus, the maximal number of frames which can be delivered is $V_I^{(m)} = (T_I - \zeta)/\zeta$. Considering the Poisson arrival model, the average number of backlogged frames at the beginning of the reserved period is

$$W = \sum_{k=\lceil V_I^{(m)} \rceil + 1}^{\infty} \left(k - \lceil V_I^{(m)} \rceil \right) \frac{(T_I \lambda)^k}{k!} e^{-T_I \lambda}. \tag{5.20}$$

During T_B, the frames stored in the buffer and those arrived during T_S are transmitted in burst mode. We have $V_B = W + (\zeta + T_S)\lambda$. Thus, the buffer is empty after T_B on average, which is consistent with the necessary condition in (5.19).

The reduced frame arrival rate for contention access is

$$\lambda' = \frac{T_Y \lambda - V_B}{T_Y}, \tag{5.21}$$

and we have $V_I = T_Y \lambda - V_B = T_Y \lambda'$.

(4) Access Time

To obtain the collision probability and average service time of a frame, we investigate the busy procedure of the tagged station, which includes backoff slots and transmission slots.

During T_A, a time slot can have two states. First, a slot may be idle if no station transmits and has the duration of δ (specified in the wireless standards). Second, a slot may contain frame transaction(s), either successful or with a collision, with the duration of $\Delta = \phi + AIFS$.

Given the condition that the tagged station is performing backoff, the probabilities for one generic slot inside T_A to be one of the two states are a_A and b_A, and

$$\begin{cases} a_A = (1 - \rho\tau)^{N-1}, \\ b_A = 1 - a_A, \end{cases} \tag{5.22}$$

where ρ is the probability for a station to be busy (i.e., performing backoff or transmitting) and τ is the probability for a station to transmit given the condition that it is busy. Then the average duration of a generic slot inside T_A is

$$S_A = a_A \delta + b_A \Delta. \tag{5.23}$$

The number of generic slots inside T_A is $\Gamma_A = \frac{T_A}{S_A}$, where $T_A = T_U - T_V - \Delta_B$ and T_V and Δ_B are calculated as follows.

(5) Vulnerable Time

If a transaction begins inside the time interval $[T_F, \ T_F + \Delta]$ before the beginning of a reserved period, it ends during the conflict time. In such a scenario, we approximate that the starting time of the transaction is uniformly distributed in $[T_F, \ T_F + \Delta]$, so the average vulnerable time is $\frac{T_F}{2}$. If no transmission occurs in the interval, the vulnerable time is T_F. The number of idle slots in $[T_F, \ T_F + \Delta]$ is $\Gamma_\Delta = \frac{\Delta}{\delta}$ and the probability of no transmission initiated in this interval is $a_A^{\Gamma_\Delta}$. The average vulnerable time is thus $a_A^{\Gamma_\Delta} T_F + \left(1 - a_A^{\Gamma_\Delta}\right) \frac{T_F}{2}$. Besides, as the arrival time of the following reserved period is uniformly distributed in the last idle slot, the slot (with the duration of $\frac{\delta}{2}$) is combined with the burst transaction slot.

Thus, by approximating T_F to $\phi + AIFS = \Delta$ for simplicity, the average length of the vulnerable time is

$$T_V = a_A^{\Gamma_\Delta} T_F + \left(1 - a_A^{\Gamma_\Delta}\right) \frac{T_F}{2} - \frac{\delta}{2} = \left(1 + a_A^{\Gamma_\Delta}\right) \frac{\Delta}{2} - \frac{\delta}{2}. \tag{5.24}$$

A slot in T_V has a duration of δ because the transactions are suppressed by the conflict avoidance. The average number of slots in T_V can be estimated by $\Gamma_V = T_V / \delta$.

(6) Reservation Period with Back-to-Back Transactions

As mentioned earlier, there are V_B frames transmitted back-to-back (in burst mode) during T_B. If V_B is smaller than one, it can be regarded as the probability that there is one frame transaction. Therefore, T_B can be obtained as

$$T_B = \begin{cases} V_B \phi_B - MIFS + SIFS + T_{ACK}, & \text{if } V_B \geq 1 \\ V_B \phi, & \text{if } V_B < 1 \end{cases} \tag{5.25}$$

where ϕ_B and ϕ are the frame transaction duration in the burst mode and in the single-frame mode, respectively, as defined in Sect. 5.1.1. For the contention-based access, we can combine the last slot before the burst transaction (as described earlier), the burst transaction, and the following $AIFS$ together as a single, busy slot. The average duration of such slots, denoted by Δ_B, equals

$$\Delta_B = \frac{\delta}{2} + T_B + AIFS. \tag{5.26}$$

The number of slots in T_B is obviously $\Gamma_B = 1$.

(7) Released Reservation Periods

Because the contention mechanism in the released period is the same as that in the contention period, we can combine them together as T_A, as shown in Fig. 5.3. The channel access behaviors for this period have been discussed in Sect. 5.1.2.

(8) Distribution of Generic Slots

During the small renewal cycle, T_U, the total number of generic slots is $\Gamma_U = \Gamma_A + \Gamma_V + \Gamma_B$. The probabilities for a slot to be in one of the three periods are, $h_A = \Gamma_A/\Gamma_U$, $h_V = \Gamma_V/\Gamma_U$, and $h_B = \Gamma_B/\Gamma_U$, respectively. The average length of a generic slot during T_U is $S_U = T_U/\Gamma_U$.

(9) Collision Probability

We consider the backoff procedure of the tagged station during T_U. If the transmission trial happens during T_A, the collision occurs if any other station also transmits in the same slot. For the transmission trial in T_V and T_B, the collision probability is one, because the backoff conflict avoidance has the same consequence as a collision (i.e., a virtual collision).

Therefore, the collision probability for the tagged station when it sends a frame in a slot during T_I is

$$p = 1 - \left[(1 - h_V - h_B)(1 - \rho\tau)^{N-1}\right]. \tag{5.27}$$

(10) Backoff Procedure, Service Time, and Station Busy Probability

After a collision or conflict, a station tries to retransmit its frame. Let $E[R]$ and $E[B]$ denote the average number of transmission slots and backoff slots experienced by a frame, respectively. Given the collision probability of p and the CW of $[0, W_k]$ for the k-th backoff stage, $E[R]$ and $E[B]$ are derived as (5.2) in Sect. 5.2.1.

When a station is busy, i.e., performing backoff or transmitting, the probability to transmit a frame in a slot equals

$$\tau = \frac{E[R]}{E[R] + E[B]}. \tag{5.28}$$

A transmission slot may have the duration of Δ (during T_A), δ (due to the transaction suppression during the vulnerable time), and Δ_B (collision with the burst transaction). Thus the average transmission slot duration is $\bar{\Delta} = h_A\Delta + h_V\delta + h_B\Delta_B$.

The contention-based service time during T_I is given by

$$\zeta \approx E[B]S_U + (E[R] - 1)\bar{\Delta} + \Delta. \tag{5.29}$$

The station busy probability, ρ, is determined by the reduced traffic level and

$$\rho = \zeta\lambda', \tag{5.30}$$

where λ' is from (5.21).

Solving the equations from (5.20) to (5.30) by the fixed-point method, we can obtain the mean values of the system parameters, such as τ, ρ, p, $E[B]$, $E[R]$, ζ, etc.

(11) Average Service Time and Throughput

The service time of the reservation-based access is approximated by the burst transaction time, Δ_B. Thus, the average service time of all frames is

$$\bar{\zeta} = \frac{V_I}{V_I + V_B}\zeta + \frac{V_B}{V_I + V_B}\Delta_B. \tag{5.31}$$

In a stable network, ignoring the limit of MAC buffer size, the per-station throughput (in bps) is

$$\eta = \left[V_B + V_I(1 - p^K)\right]\frac{L_P}{T_Y}, \tag{5.32}$$

where L_P is the payload size of a frame in bits.

5.3 Performance Evaluation

(1) Simulation Setting

To verify the accuracy of the analysis, extensive simulations have been conducted using a discrete event-driven simulator. All the numerical results reported here are obtained based on the WiMedia ECMA-368 standard [49]. As shown in Fig. 5.1, T_S is the average duration of periods reserved by DRP. T_C is the average duration of the contention-based periods between two consecutive DRP periods, where the PCA protocol is adopted. If there are D DRP periods in one superframe, we have $T_C = \frac{T_{SF}}{D} - T_S$, where $T_{SF} = 65{,}536\,\mu s$ is the superframe length.

The PHY and MAC parameters defined in the ECMA-368 standard are listed in Table 5.1. The PHY-layer data rate is 480 Mbps and the payload size of a frame is 1000 bytes. Since we simulate the transfer of video streams, each frame is encapsulated in IP/UDP/RTP. The overhead sizes of IP, UDP, and RTP are 20, 8, and 12 bytes, respectively. More detailed description of ECMA-368 is presented in Sect. 2.3. We assume that the frame arrival of each flow follows a Poisson process.

Table 5.1 Parameters used for hybrid MAC performance evaluation

Parameter	Value	Parameter	Value	Parameter	Value
AIFS	$28\,\mu s$	SIFS	$10\,\mu s$	GT	$12\,\mu s$
T_{DATA}	$31.9\,\mu s$	T_{ACK}	$13.1\,\mu s$	δ	$9\,\mu s$
T_{DRP}	$256\,\mu s$	K	7	CW_1	7

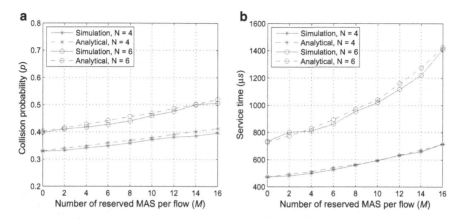

Fig. 5.4 Performance of saturated PCA stations with different M (**a**) Collision probability. (**b**) Service time

Note that, in the following figures, the collision probability of a transmission using the *backoff strategy* includes both the real collisions with other transmission and the virtual collisions with DRP periods.

(2) Hard-Reservation, Saturated Stations, Backoff Strategy

We assume that there are N PCA and N DRP stations. Each DRP station reserves M DRP periods and each period contains one MAS ($256\,\mu s$). Thus there are totally $D = NM$ DRP periods evenly distributed inside one superframe and $T_S = 256\,\mu s$.

For the saturated case, each station is backlogged with frames to send. Figure 5.4 shows the overall collision probability and the average frame service time, with $N = 4$ and $N = 6$ flows, using the backoff conflict avoidance strategy. We can see a good agreement between the analytical and simulation results.

The subfigures indicate how the performance of contention-based access changes when more channel time is reserved (as M increases). In Fig. 5.4a, we observe a slow increase of the overall collision probability. With a larger M, the DRP periods occur more frequently, which increases the virtual collision probability. However, due to virtual collisions, the average CW is larger than that of PCA only MAC. Then the transmission probability in a given time slot is reduced and the real collisions among the PCA stations are actually reduced. Therefore, increasing M does not have a significant impact on the overall collision probability.

However, as shown in Fig. 5.4b, the service time increases fast with respect to M. This is expected because the presence of DRP periods enlarges the backoff slots considerably.

(3) Hard-Reservation, Unsaturated Stations, Backoff Strategy

For the unsaturated case, we assume that the frame arrival events at each PCA station follow a Poisson process with the average inter-arrival time of $\mu = 1000\,\mu s$. The reservations from the N DRP stations are the same as above. Figure 5.5 shows

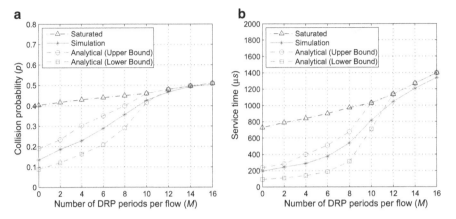

Fig. 5.5 Performance of unsaturated PCA stations with different M (**a**) Collision probability. (**b**) Service time

the collision probability and average service time with $N = 6$ flows. The results of the saturated cases are also plotted for comparison.

When M increases, the collision probability and service time increase accordingly, similar to the saturated case. Note that when $M \leq 10$, the service time is smaller than the arrival interval $\mu = 1000\,\mu s$ (as shown in Fig. 5.5b) and thus the stations are unsaturated. We can see that the analytical models can give valid lower and upper bounds. However, when $M \geq 12$, the service time becomes larger than μ, which indicates that the stations have become saturated. Both bounds converge to the results of saturated stations, as expected due to the asymptotic property given in Sect. 5.2.2. Also we can see that the convergence of the lower and upper bound models correctly predicts the transition from unsaturated to saturated status (when μ is smaller than the average frame service time).

Note that for $M \geq 12$, although the stations have become saturated, the average frame service time is only slightly larger than the inter-arrival interval μ, and there still exist some moments that a station is idle due to the burstiness of the traffic. Therefore, the simulation results are slightly lower than the analytical results for a real saturated station (which is always busy).

(4) Comparison of Backoff and Hold-on Strategies for Hard-Reservation

To compare the performance of the two conflict avoidance strategies, we present the simulation results of $N = 6$ and $N = 10$ saturated stations, because the saturation throughput indicates the limit of the network capacity. The reservations of the DRP periods are the same as above.

Figure 5.6a shows that the collision probability (those collisions resulting in backoff) using the backoff strategy is slightly higher than that of the hold-on strategy. The difference is because of the following scenario: the tagged station obtains TXOPs during the vulnerable time T_V. According to the backoff strategy,

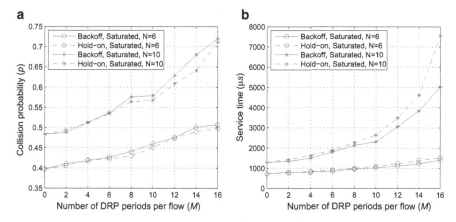

Fig. 5.6 Performance of the two conflict avoidance strategies with different M (**a**) Collision probability. (**b**) Service time

a virtual collision happens definitely and the station invokes a new stage of backoff. However, if the hold-on strategy is used, this station holds on and may successfully transmit in the slot immediately following the DRP period (i.e., no other station obtains TXOPs during T_V). In this case, there is no collision. This scenario is actually presented analytically by comparing (5.10) and (5.11). Obviously, p calculated in (5.10) (backoff strategy) is larger than that in (5.11) (hold-on strategy). However, given that the stations are saturated, the probability of this no-collision case for the hold-on strategy is quite small. Consequently, in the hold-on strategy, a TXOP obtained during T_V will experience a collision after the DRP period with a high probability. Therefore, the difference of the collision probabilities of the two strategies is small.

However, the average frame service time for the backoff strategy is smaller, as shown in Fig. 5.6b. This is because, with the backoff strategy, if a station obtains a TXOP during T_V, it immediately starts the next backoff stage. So after DRP, the station may finish its backoff earlier to transmit. But with the hold-on strategy, a station will delay its transmission until the slot after DRP, where it also has a high probability to have a collision with other PCA stations. Thus, the hold-on strategy may waste the slots in the vulnerable time for backoff. In addition, using the hold-on strategy, each collision results in wasted channel time of frame transaction Δ, but with the backoff strategy, a virtual collision does not waste additional channel time.

Note that, when M increases, the percentage of the vulnerable time in the total channel time increases. Hence, the service time difference between the two strategies becomes larger. In summary, for saturated stations, the backoff strategy can lead to a higher throughput. This has also been discussed in [112].

(5) Soft-Reservation, Unsaturated Stations, Backoff Strategy

Figure 5.7 shows the overall collision probability and the average frame service time of the hybrid MAC using soft reservation, single buffer, and the backoff conflict

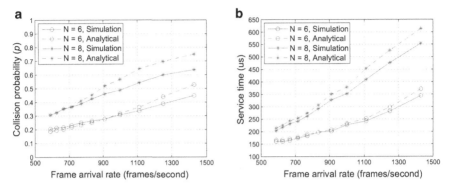

Fig. 5.7 Performance vs. frame arrival rate for soft reservation (**a**) Collision probability.
(**b**) Service time

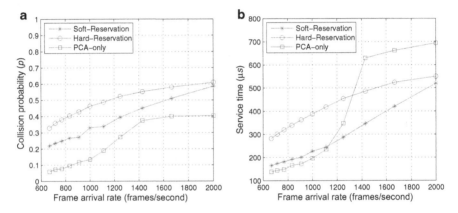

Fig. 5.8 Performance vs. frame arrival rate for the three MAC protocols (**a**) Collision probability.
(**b**) Service time

avoidance strategy. We set $N = 4$ and $N = 6$ flows. The simulation results match reasonably well with the analytical ones.

(6) Comparison of Three MAC Protocols

Figure 5.8 shows the comparison of the three MAC protocols, i.e., the soft-reservation, hard-reservation, and conventional contention-only MAC, with $N = 6$ stations.

First, the collision probability and average frame service time with hard reservation are always larger than those with soft reservation. It is expected because, by using soft reservation, the unused DRP periods can be released and used by the others. Furthermore, when the frame arrival rate is large (≥ 2000 frames/s), their performance converges. Because when the load increases, more frames are

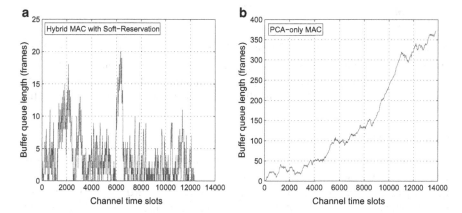

Fig. 5.9 Queue length evolution of a sender's buffer (**a**) Hybrid MAC. (**b**) Contention-based MAC

backlogged in the buffers. Thus, the burst size is increased and less time is released, so soft reservation performs similar to hard reservation.

Second, the collision probability of the hybrid MAC is always larger than that of the contention-only MAC. Note that, in the figures, the collision probability includes both the real collisions between the PCA transactions and the conflict (virtual collisions) with the DRP periods. The presence of the reservation (i.e., DRP periods) introduces conflicts for the contention-based access in hybrid MAC and therefore increases the collision probability.

Third, the service time of the contention-only MAC is smaller than that of soft reservation when the traffic load is light (e.g., 1000 frames/s). This is because there is almost no frame backlogged in the buffers when the reserved periods arrive. The reservation is not utilized but just introduces conflict, resulting in a larger collision probability and service time. However, when the traffic load is heavy, the service time of the contention-only MAC increases quickly, because the network tends to be saturated. But the reservation in hybrid MAC can be fully utilized to transfer the traffic in the burst mode efficiently, leading to a much smaller service time.

Fourth, Fig. 5.9 plots the queue length evolution of a sender's buffer using the contention-based MAC and the hybrid MAC with soft reservation. The frame arrival rate is 2000 frames/s. We can see that the queue length of the contention-based MAC (Fig. 5.9b) keeps increasing (close to 400 frames). Hence the network is overloaded (saturated) and cannot work. However, the queue length of the hybrid MAC (Fig. 5.9a) is still limited (less than 20 frames), which shows that the network is heavily loaded but can still support the traffic and is stable. This further illustrates the advantages of the hybrid MAC, in particular in supporting heavy traffic.

In summary, the hybrid MAC with soft reservation can accommodate a higher traffic arrival rate, so its admission region is larger than the contention-only MAC and the hybrid MAC with hard reservation.

5.4 Case Study: Supporting HDTV

In this section, we study the case of using the hybrid MAC to support HDTV [155]. We first outline the simulation setup, including the simulation scenarios and parameters, then present the analysis and simulation results for video streaming over hybrid DRP/PCA MAC, and finally give the admission region considering the IPTV-like applications.

(1) Simulation Setup

We adopt a trace-driven simulation strategy to demonstrate the effectiveness of the hybrid MAC. In this section, the sample HD video stream is "From Mars to China" with a resolution of 1920×1080. The video is H.264/MPEG-4 AVC encoded with quantization parameters of 28, 28, and 30 for I, P, and B frames, respectively. The video is available at http://trace.eas.asu.edu/h264/mars/. In this sample video, the maximum video frame size is $326, 905$ bytes and the average frame size is $20, 209$ bytes with high burstiness, as shown in Fig. 1.1. If the video frame size is 1000 bytes, the average video data rate is 621.486 frames per second for each flow.

We employ a commonly-used network simulator, NS-2, and have extended TKN's IEEE 802.11e code to simulate WiMedia MAC protocols. The superframe duration is 65.536 ms for 256 MASs of $256 \mu s$ each. The transmitter needs to ensure that the whole frame transaction including acknowledgment should finish at least one *SIFS* plus one guard time before the PCA period transitions into DRP, or vice versa. Otherwise, it will hold on the frame transaction completely until the channel is available for PCA again. These behaviors due to the hybrid DRP/PCA MAC have been captured closely in our simulation.

In WiMedia PCA, for video traffic, aCW_{min} is 7 and the retry limit is 7, the AIFSN is 2, the slot time δ is $9 \mu s$, and the SIFS is $10 \mu s$, so the video's AIFS is $28 \mu s$. The guard time is $12 \mu s$. During the PCA periods, a frame transaction includes the time to transmit a data frame and an acknowledgment, as well as SIFS and AIFS. During the DRP periods, since the reservation owner has the exclusive access to the channel, a frame transaction includes the time to transmit the frame and acknowledgment, as well as at most two SIFSs. The video frames are encapsulated in RTP, UDP, IP frames, and WiMedia LLC frames, with a total overhead of 56 bytes, before going to the PLCP layer.

The WiMedia data rate of 480 Mbps is used. According to WiMedia's standard, PLCP preamble and header are also considered in our simulation and transmitted at a lower data rate, while the video payload including the upper-layer headers is transmitted at the given data rate (480 Mbps). For a video frame of 1000 bytes, it takes $31.875 \mu s$ to transmit the entire PLCP frame, and the MAC-layer acknowledgment takes $13.125 \mu s$. We ignore the propagation delay due to the short range in UWB networks.

We adopt the dual buffer architecture, as depicted in Fig. 5.2b. The R-Buffer stores the video frames to be transmitted in the following DRP MASs reserved by the station. Given the maximal number of frames that can be served in one MAS (denoted by F_{MAS}) and the number of reserved slots, we can determine the R-Buffer size, such that the time to transmit a full buffer of frames does not exceed the video delay jitter bound allowed in the wireless network. When the R-Buffer is full, the excess video frames will be put into the C-Buffer, where they can compete to access the channel during the PCA periods.

(2) DRP/PCA Traffic Breakdown

Here we present the results when a certain number (M) of MASs are reserved for each video flow, and the remaining is available for PCA among all flows. Thus, the hybrid DRP/PCA approach has $M > 0$ and the PCA-only approach has $M = 0$.

We pre-process the video trace to determine the number of video frames transmitted through DRP and PCA. If we consider I-ACK for DRP frames, which are separated by SIFS between frame transactions due to exclusive access, a total frame transaction lasts $31.875 + 10 + 13.125 + 10 = 65\,\mu s$. For a reserved MAS, it can accommodate $\lfloor(256-12)/65\rfloor = 3$ frames due to the guard time. If we consider B-ACK, one MAS can accommodate $\lfloor(256-12-10-13.125)/(31.875+10)\rfloor = 5$ frames. If we consider both B-ACK and burst transmission with MIFS, one MAS can accommodate $\lfloor(256 - 12 - 10 - 13.125 - 10 + 1.875)/(31.875 + 1.875)\rfloor = 6$ frames. Even with burst PLCP preamble in the physical layer, at most 6 frames can be accommodated in one MAS.

As shown in Fig. 5.10, the percentage of the remaining PCA traffic after DRP reservation strictly decreases as the number of reserved slots increases. However, the percentage reduction is much slower than the increase of DRP slots. This is due to the burstiness of video traffic, which cannot fully utilize all reserved slots in each superframe. At the beginning, one more reserved MAS can reduce the remaining PCA traffic greatly, especially when one MAS can accommodate 6 frames with B-ACK and MIFS. After a certain number of MASs are reserved, an additional reserved MAS only has a marginal benefit for I frames, since there is no remaining traffic for P and particularly B frames any further. For example, with B-ACK and MIFS (which is the most efficient way to use DRP), even with 16 MASs reserved in one superframe, more than 10 % of the total traffic (mainly the I frames) still has to go through PCA. This indicates that a suitable number of reserved slots should be chosen carefully. For illustration purposes, we use B-ACK and MIFS for the simulations presented in this chapter, i.e., 6 video frames going through a DRP MAS.

(3) Frame Service Time

Figure 5.11 shows the average frame service time obtained from both analysis and simulation for PCA frames. In this figure, there are 8, 10, and 12 concurrent video flows. As being expected, when M is small, the service time decreases with regard to M, and the trend is reversed with a large M. This is because, a small number of reserved MASs can be efficiently utilized and the number of frames left for contention in PCA is greatly reduced. With less contention, the overhead (collisions and backoff) in the PCA periods can be reduced as well. When M is large,

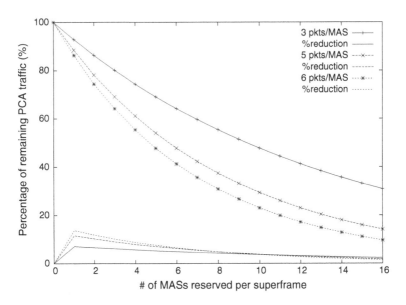

Fig. 5.10 The percentage of the remaining PCA traffic after DRP reservation

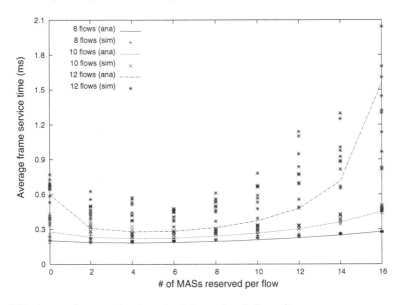

Fig. 5.11 Average frame service time: simulation and analysis results

since the duration of a MAS for DRP (256 μs) is much longer than that of a fixed contention slot for PCA (9 μs), frequent interruption by the DRP periods will bring up the service time in PCA. In addition, with a large M, the DRP reserved MASs are not efficiently utilized, but the remaining channel time for PCA is significantly

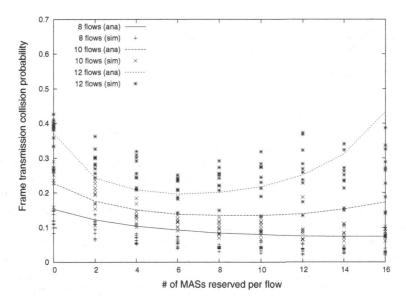

Fig. 5.12 Frame collision probability: simulation and analysis results

reduced, which results in a higher contention level and longer service time. The frame service time "bounce-back" behavior shows that there is an optimal number of MASs reserved for each video flow.

(4) Frame Collision Probability

Figure 5.12 shows the frame transmission collision probability obtained from both the analysis and simulation for the PCA frames. As shown in the figure, when there is a small number of MASs reserved for DRP, the frame collision probability for PCA frames is actually reduced, due to fewer active stations contending for the channel during PCA periods. However, when the number of MASs reserved for DRP increases, more PCA backoff processes are interrupted by the DRP periods. After the DRP reserved periods, the PCA frames may collide due to the "pre-transmission" withhold when the time before the DRP periods is not enough to finish the frame transaction. In addition, when more MASs are reserved for DRP, fewer will be available for PCA, which leads to a higher chance of contention overall. This trend is more obvious when the number of video flows is large (e.g., 12 flows). For 8 flows, the same "bounce-back" behavior is expected with the slow reduction in collision probability and the increase in frame service time in Fig. 5.11. The optimal number of reserved MASs in a superframe (M) to minimize the frame service time and collision probability depends on the number of flows (N).

In the analytical framework presented in Sect. 5.2, we only consider average traffic arrival rate and assume the independence of competing flows, which should be general enough to investigate the performance of other traffic types with different bursty levels. Besides using the video traces, we also use Poisson traffic in simulation for verification. Simulation results with both video traces and Poisson traffic validate the correctness and wide applicability of the analytical models presented in Sect. 5.2.

(5) Admission Region

The admission region is determined by ensuring both frame loss rate (FLR) and delay jitter for video streams to meet their QoS requirement. For IPTV-like applications, FLR should be less than 10^{-4} and the delay jitter should be less than 100 ms. FLR can be obtained from the collision probability and the maximum video frame jitter is due to the queuing delay of the largest video frames over the wireless networks. Our analytical and simulation results show that, for the PCA-only and hybrid MAC, the delay jitter constraint for HD video is tighter than the FLR constraint due to collisions. Therefore, the number of video flows that can be supported is mainly determined by the frame service time of video frames.

For the PCA-only and hybrid MAC, we use the frame service time for the largest video frame (around 327 KB) as the admission criterion. It is estimated as the number of frames being transmitted in PCA periods times the frame service time of PCA. For the DRP-only MAC, we can calculate the MASs needed for each video and then determine the admission region accordingly.

To ensure that the maximum delay jitter is less than two or three video-frame durations (i.e., 66.67 ms and 100 ms, respectively), only 5 and 7 video streams can be supported with the DRP-only MAC, as we need to over-reserve significantly for the bursty video traffic. As shown in Fig. 5.13, for the PCA-only MAC (i.e., the number of MASs reserved per flow is 0), at most 8 and 10 flows can be supported to meet the two criteria of maximum video frame jitter. For the hybrid MAC and when each DRP MAS can send 6 frames, we can support 10 and 13 flows if we reserve 6 MASs for each flow, which outperforms both the PCA-only and DRP-only MAC considerably. This admission region comparison clearly shows the tradeoff between contention and reservation-based medium access control mechanisms, and the way to strike a better balance between them.

5.5   Summary

In this chapter, we have introduced the hybrid MAC protocols to better support video streaming over wireless networks and studied the QoS performance and admission region. Furthermore, we have discussed two conflict avoidance strategies, and the simulation results show that the backoff strategy can achieve a higher throughput when the channel time is reserved frequently. Two buffering architectures are also discussed and it is shown that the dual-buffer can provide a considerably better performance for hard-reservation approach, thanks to the higher reservation utilization and lower contention level. We have also discussed the hybrid MAC using soft-reservation, where the unused reserved time can be released implicitly and accessed by other stations through contention.

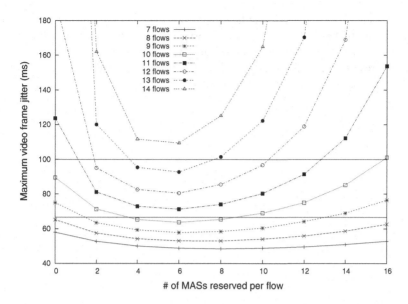

Fig. 5.13 Admission region determined by the maximum video frame jitter

An accurate analytical model based on the mean value analysis for the hard-reservation with saturated stations has been presented first. The collision probability and the average service time of a frame are obtained. For the unsaturated case, we have presented tight lower and upper bounds which converge when the stations become saturated. The analytical framework for the hybrid MAC using soft-reservation has also been presented.

The hybrid MAC based on the WiMedia ECMA-368 standard has been simulated to validate the analysis and compare the performance between the conventional contention-only MAC and the hybrid MAC. Analysis and simulation results show that the soft reservation has the best performance and reaches high capacity when the traffic load is relatively high. In addition, using the WiMedia UWB WPAN as an example, a case study with a real video trace and the NS-2 simulator has been conducted. The simulation results validate the analytical model of the hybrid MAC and also show the admission region in supporting HDTV streaming flows.

Both analytical and simulation results demonstrate that the hybrid MAC is desired for high-quality video streaming. This is because, if we reserve a portion of channel time for each video flow, the reserved periods can be efficiently utilized and at the same time the collision and service time in contention periods are significantly reduced. Thus, better QoS and even more video flows can be supported.

As many wireless standards adopt the interleaved reservation and contention MAC protocols, the hybrid MAC approach reported in this chapter is anticipated to be a key enabling technology to bridge the gap. There are many open issues beckoning for further research, for instance, how to optimize the reservation duration and pattern, considering heterogeneous data and multimedia traffic; how to make the tradeoff of hard and soft reservation to further improve the network efficiency; how to fine-tune the contention access protocols at the presence of frequent reservation periods; how to flexibly utilize the hybrid MAC approach to deal with channel fading and interference, etc.

References

1. Aad, I., Castelluccia, C.: Differentiation mechanisms for IEEE 802.11. In: Proceedings of IEEE/ACM International Conference on Computer Communications (INFOCOM), Anchorage, AK, pp. 209–218 (2001)
2. Akyildiz, I.F., McNair, J., Martorell, L.C., Puigjaner, R., Yesha, Y.: Medium access control protocols for multimedia traffic in wireless networks. IEEE Netw. **13**(4), 39–47 (1999)
3. Amendment of parts 2, 15, and 97 of the commission's rules to permit use of radio frequencies above 40 GHz for new radio applications. Technical Report FCC 95-499, Federal Communications Commission (FCC) (1995)
4. Amitay, N., Greenstein, L.J.: Resource auction multiple access (RAMA) in the cellular environment. IEEE Trans. Veh. Technol. **43**(4), 1101–1111 (1994)
5. Assi, C.M., Agarwal, A., Liu, Y.: Enhanced per-flow admission control and QoS provisioning in IEEE 802.11e wireless LANs. IEEE Trans. Veh. Technol. **57**(2), 1077–1088 (2008)
6. Barry, M.G., Campbell, A.T., Veres, A.: Distributed control algorithms for service differentiation in wireless packet networks. In: Proceedings of IEEE/ACM International Conference on Computer Communications (INFOCOM), Anchorage, AK, pp. 582–590 (2001)
7. Batra, A., Balakrishnan, J., Aiello, G.R., Foerster, J.R., Dabak, A.: Design of a multiband OFDM system for realistic UWB channel environments. IEEE Trans. Microwave Theory Tech. **52**(9), 2123–2138 (2004)
8. Battiti, R., Li, B.: Supporting service differentiation with enhancements of the IEEE 802.11 MAC protocol: models and analysis. Technical Report, University of Trento, Trento (2003)
9. Baykas, T., Sum, C.S., Lan, Z., Wang, J., Rahman, M.A., Harada, H., Kato, S.: IEEE 802.15.3c: the first IEEE wireless standard for data rates over 1 Gb/s. IEEE Commun. Mag. **49**(7), 114–121 (2011)
10. Benveniste, M.: TCMA proposed draft text. Technical Report Doc. 802.11-01/117r2, IEEE Working Group (2001)
11. Beshai, M.: The poissonian-spectrum method for treating a loss system serving non-poissonian multi-bit-rate traffic. In: Proceedings of IEEE/ACM International Conference on Computer Communications (INFOCOM), Ottawa, ON, pp. 1010–1019 (1989)
12. Bianchi, G.: Performance analysis of the IEEE 802.11 distributed coordination function. IEEE J. Sel. Areas Commun. **18**(3), 535–547 (2000)
13. Bianchi, G., Tinnirello, I.: Analysis of priority mechanisms based on differentiated inter frame spacing in CSMA/CA. In: Proceedings of IEEE Vehicular Technology Conference (VTC-Fall), Jeju Island, pp. 1401–1405 (2003)

© The Author(s) 2017

R. Zhang et al., *Resource Management for Multimedia Services in High Data Rate Wireless Networks*, SpringerBriefs in Electrical and Computer Engineering, DOI 10.1007/978-1-4939-6719-3

14. Bianchi, G., Tinnirello, I.: Remarks on IEEE 802.11 DCF performance analysis. IEEE Commun. Lett. **9**(8), 765–767 (2005)
15. Bluetooth core version 4.0 specification. Technical Report V4.0, Bluetooth special interest group (SIG) (2010). https://www.bluetooth.org/Technical/Specifications/adopted.htm
16. Bluetooth core version 4.1 specification. Technical Report V4.0, Bluetooth special interest group (SIG) (2013). https://www.bluetooth.org/Technical/Specifications/adopted.htm
17. Boulis, A., Smith, D., Miniutti, D., Libman, L., Tselishchev, Y.: Challenges in body area networks for healthcare: the MAC. IEEE Commun. Mag. **50**(5), 100–106 (2012)
18. Cai, X.L., Shen, X., Cai, L., Mark, J.W., Xiao, Y.: Voice capacity analysis of WLAN with unbalanced traffic. IEEE Trans. Veh. Technol. **55**(3), 752–761 (2006)
19. Cao, H., Leung, V., Chow, C., Chan, H.: Enabling technologies for wireless body area networks: A survey and outlook. IEEE Commun. Mag. **47**(12), 84–93 (2009)
20. Chatzimisios, P., Boucouvalas, A., Vitsas, V.: IEEE 802.11 packet delay a finite retry limit analysis. In: Proceedings of IEEE Global Communications Conference (GLOBECOM), San Fransisco, CA, pp. 950–954 (2003)
21. Chen, D., Gu, D., Zhang, J.: Supporting real-time traffic with QoS in IEEE 802.11e based home networks. In: Proceedings of IEEE Consumer Communications and Networking Conference (CCNC), Las Vegas, NV, pp. 205–209 (2004)
22. Chen, X., Zhai, H., Tian, X., Fang, Y.: Supporting QoS in IEEE 802.11e wireless LANs. IEEE Trans. Wirel. Commun. **5**(8), 2217–2227 (2006)
23. Cheng, H.T., Zhuang, W.: Novel packet-level resource allocation with effective QoS provisioning for wireless mesh networks. IEEE Trans. Veh. Technol. **8**(2), 694–700 (2009)
24. Cheng, Y., Ling, X., Song, W., Cai, L.X., Zhuang, W., Shen, X.: A cross-layer approach for WLAN voice capacity planning. IEEE J. Sel. Areas Commun. **25**(4), 678–688 (2007)
25. Chesson, G., et al.: EDCF proposed draft text. Technical Report Doc. 802.11-01/131r1, IEEE Working Group (2001)
26. Chou, C.T., Sai-Shankar, N., Shin, K.G.: Achieving per-stream QoS with distributed airtime allocation and admission control in IEEE 802.11e wireless LANs. In: Proceedings of IEEE/ACM International Conference on Computer Communications (INFOCOM), Miami, FL, pp. 1584–1595 (2005)
27. Chung, W.S., Un, C.K.: Collision resolution algorithm for M-priority users. IEE Proc. Commun. **142**(3), 151–157 (1995)
28. Cisco visual networking index: flobal mobile data traffic forecast update, 2012–2017. Technical Report, Cisco (2013)
29. Daneshi, M.: Distributed reservation algorithms for video streaming over WiMedia UWB networks. Master's thesis, University of Victoria, BC (2009)
30. Daneshi, M., Pan, J., Ganti, S.: Distributed reservation algorithms for video streaming over UWB-based home networks. In: Proceedings of IEEE Consumer Communications and Networking Conference (CCNC), Las Vegas, NV, pp. 1–6 (2010)
31. Daneshi, M., Pan, J., Ganti, S.: Towards an efficient reservation algorithm for distributed reservation protocols. In: Proceedings of IEEE/ACM International Conference on Computer Communications (INFOCOM), San Diego, CA, pp. 1855–1863 (2010)
32. Delbrouck, L.: A unified approximate evaluation of congestion functions for smooth and peaky traffics. IEEE Trans. Commun. **29**(2), 85–91 (1981)
33. Deng, D.J., Yen, H.C.: Quality-of-service provisioning system for multimedia transmission in IEEE 802.11 wireless LANs. IEEE J. Sel. Areas Commun. **23**(6), 1240–1252 (2005)
34. Djukic, P., Valaee, S.: Distributed link scheduling for TDMA mesh networks. In: Proceedings of IEEE International Conference on Communications (ICC), Glasgow, pp. 3823–3828 (2007)
35. Duan, C., Pekhteryev, G., Fang, J., Nakache, Y., Zhang, J., Tajima, K., Nishioka, Y., Hirai, H.: Transmitting multiple HD video streams over UWB links. In: Proceedings of IEEE Consumer Communications and Networking Conference (CCNC), Las Vegas, NV, pp. 691–695 (2006)
36. Elnoubi, S., Alsayh, A.M.: A packet reservation multiple access (PRMA)-based algorithm for multimedia wireless system. IEEE Trans. Veh. Technol. **53**(1), 215–222 (2004)

37. End-user multimedia QoS categories. Technical Report G.1010, International Telecommunication Union-Telecommunication Standardization Sector (ITU-T) (2001)
38. Engelstad, P., Osterbo, O.: Analysis of the total delay of IEEE 802.11e EDCA and 802.11 DCF. In: Proceedings of IEEE International Conference on Communications (ICC), Istanbul, pp. 552–559 (2006)
39. Falconer, D.D., Adachi, F., Gudmundson, B.: Time division multiple access methods for wireless personal communications. IEEE Commun. Mag. **33**(1), 50–57 (1995)
40. First report and order in the matter of revision of part 15 of the commission's rules regarding ultra-wideband transmission systems. Technical Report FCC 02-48, Federal Communications Commission (FCC) (2002)
41. Foh, C., Zukerman, M., Tantra, J.: A markovian framework for performance evaluation of IEEE 802.11. IEEE Trans. Wirel. Commun. **6**(4), 1276–1285 (2007)
42. Frigon, J.F., Leung, V.C.M., Chan Bun Chan, H.: Dynamic reservation TDMA protocol for wireless ATM networks. IEEE J. Sel. Areas Commun. **19**(2), 370–383 (2001)
43. Garetto, M., Chiasserini, C.F.: Performance analysis of the 802.11 distributed coordination function under sporadic traffic. In: Proceedings of 4th International Networking Conference, Waterloo, pp. 1–12 (2005)
44. Generic criteria for version 1.0 wireless access communications system (WACS). Technical Report TA-NWT-001313, Bellcore Technical Advisory (1992)
45. Glabowski, M., Kubasik, K., Stasiak, M.: Modelling of systems with overflow multi-rate traffic and finite number of traffic sources. In: Proceedings of International Symposium on Communication Systems, Networks and Digital Signal Processing (CSNDSP), Graz, pp. 196–199 (2008)
46. Goodman, D.: Trends in cellular and cordless communications. IEEE Commun. Mag. **29**(6), 31–40 (1991)
47. Hall, P.S., Hao, Y.: Antennas and Propagation for Body-Centric Wireless Communications. Artech House, Boston (2006)
48. He, J., Zheng, L., Yang, Z., Chou, C.T.: Performance analysis and service differentiation in IEEE 802.11 WLAN. In: Proceedings of IEEE Conference on Local Computer Networks (LCN), Bonn, pp. 691–697 (2003)
49. High rate ultra wideband PHY and MAC standard. Technical Report ECMA-368, WiMedia Alliance (2005). http://www.ecma-international.org/publications/standards/Ecma-368.htm
50. Hossain, E., Bhargava, V.K.: A centralized TDMA-based scheme for fair bandwidth allocation in wireless IP networks. IEEE J. Sel. Areas Commun. **19**(11), 2201–2214 (2001)
51. Howarth, J.A., et al.: Towards a 60 GHz gigabit System-On-Chip. In: Proceedings of Wireless World Research Forum Meeting (WWRF), Shanghai (2006)
52. Huang, Y.K., Pang, A.C., Hsiu, P.C., Zhuang, W., Liu, P.: Distributed throughput optimization for ZigBee cluster-tree networks. IEEE Trans. Parallel Distrib. Syst. **23**(3), 513–520 (2012)
53. Hui, J., Devetsikiotis, M.: Designing improved MAC packet scheduler for 802.11e WLAN. In: Proceedings of IEEE Global Communications Conference (GLOBECOM), San Francisco, CA, pp. 184–189 (2003)
54. IEEE standard for wireless medium access control (MAC) and physical layer (PHY) specifications: medium access control (MAC) quality of service enhancements. Technical Report IEEE 802.11e-2005, IEEE Standards Association - Wireless LAN Working Group (2005)
55. Jiang, H., Wang, P., Zhuang, W., Shen, X.: An interference aware distributed resource management scheme for CDMA-based wireless mesh backbone. IEEE Trans. Wirel. Commun. **6**(12), 4558–4567 (2007)
56. Kato, S., Harada, H., Funada, R., Baykas, T., Sum, C.S., Wang, J., Rahman, M.A.: Single carrier transmission for multi-gigabit 60-GHz WPAN systems. IEEE J. Sel. Areas Commun. **27**(8), 1466–1478 (2009)
57. Kumar, A., Altman, E., Miorandi, D., Goyal, M.: New insights from a fixed point analysis of single cell IEEE 802.11 WLANs. IEEE/ACM Trans. Netw. **15**(3), 588–601 (2007)

58. Lee, J.G., Corson, M.S.: The performance of an "imbedded" Aloha protocol in wireless networks. In: Proceedings of IEEE International Symposium on Personal, Indoor and Mobile Radio Communications (PIMRC), Taipei, vol. 2, pp. 377–381 (1996)

59. Ling, X.: Performance analysis of distributed MAC protocols for wireless networks. Ph.D. dissertation, University of Waterloo, Ontario (2007)

60. Ling, X., Liu, K.H., Cheng, Y., Shen, X.: A novel performance model for distributed prioritized MAC protocols. In: Proceedings of IEEE Global Communications Conference (GLOBECOM), Washington, DC, pp. 4692–4696 (2007)

61. Ling, X., Cheng, Y., Mark, J.W., Shen, X.: A renewal theory based analytical model for the contention access period of IEEE 802.15.4 MAC. IEEE Trans. Wirel. Commun. **7**(6), 2340–2349 (2008)

62. Liu, K.H., Shen, X., Zhang, R., Cai, L.: Delay analysis of distributed reservation protocol with UWB shadowing channel for WPAN. In: Proceedings of IEEE International Conference on Communications (ICC), Beijing, pp. 2769–2774 (2008)

63. Liu, K.H., Ling, X., Shen, X., Mark, J.W.: Performance analysis of prioritized MAC in UWB WPAN with bursty multimedia traffic. IEEE Trans. Veh. Technol. **57**(4), 2462–2473 (2008)

64. Liu, K.H., Shen, X., Zhang, R., Cai, L.: Performance analysis of distributed reservation protocol for UWB-based WPAN. IEEE Trans. Veh. Technol. **58**(2), 902–913 (2009)

65. Local and metropolitan area networks - wireless LAN medium access control (MAC) and physical layer (PHY) specifications. Technical Report IEEE 802.11-1997, IEEE Standards Association - LAN/MAN Standards Committee (1997)

66. Ma, X., Refai, H.H.: Analysis of sliding frame R-ALOHA protocol for real-time distributed wireless networks. Wirel. Netw. **15**(8), 1102–1112 (2009)

67. Malone, D., Dangerfield, I., Leith, D.: Verification of common 802.11 MAC model assumptions. In: Proceedings of Passive and Active Measurement Conference (PAM), Louvain-la-Neuve, pp. 63–72 (2007)

68. Matrawy, A., Lambadaris, I., Huang, C.: MPEG4 traffic modeling using the transform expand sample methodology. In: Proceedings of IEEE International Workshop on Networked Appliances (IWNA), Liverpool, pp. 249–256 (2002)

69. MBOA wireless medium access control (MAC) specification for high rate wireless personal area networks (WPANS). Technical Report MBOA MAC Specification Draft 0.65, Multiband OFDM Alliance (2004). http://www.multibandofdm.org/

70. Network simulator 2 (NS-2): version 2.33 (2008)

71. One-way transmission time. Technical Report G.114, International Telecommunication Union-Telecommunication Standardization Sector (ITU-T) (2003)

72. Papantoni-Kazakos, T., Likhanov, N.B., Tsybakov, B.: A protocol for random multiple access of packets with mixed priorities in wireless networks. IEEE J. Sel. Areas Commun. **13**(7), 1324–1331 (1995)

73. Part 11: Wireless LAN medium access control (MAC) and physical layer (PHY) specifications: high speed physical layer in the 5 GHz band. Technical Report IEEE 802.11a-1999, IEEE Standards Association - Wireless LAN Working Group (1999)

74. Part 11: Wireless LAN medium access control (MAC) and physical layer (PHY) specifications: higher speed physical layer (PHY) extension in the 2.4 GHz band. Technical Report IEEE 802.11b-1999, IEEE Standards Association - Wireless LAN Working Group (1999)

75. Part 11: Wireless LAN medium access control (MAC) and physical layer (PHY) specifications: further higher data rate extension in the 2.4 GHz band. Technical Report IEEE 802.11g-2003, IEEE Standards Association - Wireless LAN Working Group (2003)

76. Part 11: Wireless LAN medium access control (MAC) and physical layer (PHY) specifications. Technical Report IEEE 802.11-2007, IEEE Standards Association - Wireless LAN Working Group (2007)

77. Part 11: Wireless LAN medium access control (MAC) and physical layer (PHY) specifications: enhancements for higher throughput. Technical Report IEEE 802.11n-2009, IEEE Standards Association - Wireless LAN Working Group (2009)

78. Part 11: Wireless LAN medium access control (MAC) and physical layer (PHY) specifications: interworking with external networks. Technical Report IEEE 802.11u-2011, IEEE Standards Association - Wireless LAN Working Group (2011)
79. Part 11: Wireless LAN medium access control (MAC) and physical layer (PHY) specifications. Technical Report IEEE 802.11-2012, IEEE Standards Association - Wireless LAN Working Group (2012)
80. Part 11: Wireless LAN medium access control (MAC) and physical layer (PHY) specifications: enhancements for very high throughput for operation in bands below 6 GHz. Technical Report IEEE 802.11ac-2013, IEEE Standards Association - Wireless LAN Working Group (2013)
81. Part 11: Wireless LAN medium access control (MAC) and physical layer (PHY) specifications: enhancements for very high throughput in the 60 GHz band. Technical Report IEEE 802.11ad-2012, IEEE Standards Association - Wireless LAN Working Group (2012)
82. Part 11: Wireless LAN medium access control (MAC) and physical layer (PHY) specifications: television white spaces (TVWS) operation. Technical Report IEEE 802.11af-2013, IEEE Standards Association - Wireless LAN Working Group (2013)
83. Part 15: Wireless medium access control (MAC) and physical layer (PHY) specifications for wireless personal area networks (WPAN). Technical Report IEEE 802.15.1-2002, IEEE Standards Association - WPAN Working Group (2002)
84. Part 15.1a: Wireless medium access control (MAC) and physical layer (PHY) specifications for wireless personal area networks (WPAN). Technical Report IEEE 802.15.1-2005, IEEE Standards Association - WPAN Working Group (2005)
85. Part 15.2: Coexistence of wireless personal area networks with other wireless devices operating in unlicensed frequency bands. Technical Report IEEE 802.15.2-2003, IEEE Standards Association - WPAN Working Group (2003)
86. Part 15.3: Wireless medium access control (MAC) and physical layer (PHY) specifications for high rate wireless personal area networks (WPAN). Technical Report IEEE 802.15.3-2003, IEEE Standards Association - WPAN Working Group (2003)
87. Part 15.3: Wireless medium access control (MAC) and physical layer (PHY) specifications for high rate wireless personal area networks (WPAN): amendment to MAC sublayer. Technical Report IEEE 802.15.3b-2005, IEEE Standards Association - WPAN Working Group (2005)
88. Part 15.3: Wireless medium access control (MAC) and physical layer (PHY) specifications for high rate wireless personal area networks (WPAN): millimeter-wave-based alternative physical layer extension. Technical Report IEEE 802.15.3c-2009, IEEE Standards Association - WPAN Working Group (2009)
89. Part 15.4: Wireless medium access control (MAC) and physical layer (PHY) specifications for low rate wireless personal area networks (WPANs). Technical Report IEEE 802.15.4-2003, IEEE Standards Association - WPAN Working Group (2003)
90. Part 15.4: Wireless medium access control (MAC) and physical layer (PHY) specifications for low rate wireless personal area networks (WPANs). Technical Report IEEE 802.15.4-2006, IEEE Standards Association - WPAN Working Group (2006)
91. Part 15.4: Wireless medium access control (MAC) and physical layer (PHY) specifications for low rate wireless personal area networks (WPANs): add alternate PHYs. Technical Report IEEE 802.15.4a-2007, IEEE Standards Association - WPAN Working Group (2007)
92. Part 15.4: Wireless medium access control (MAC) and physical layer (PHY) specifications for low rate wireless personal area networks (WPANs): alternative physical layer extension to support one or more of the Chinese 314–316 MHz, 430–434 MHz, and 779–787 MHz bands. Technical Report IEEE 802.15.4c-2009, IEEE Standards Association - WPAN Working Group (2009)
93. Part 15.4: Wireless medium access control (MAC) and physical layer (PHY) specifications for low rate wireless personal area networks (WPANs): alternative physical layer extension to support the Japanese 950 MHz bands. Technical Report IEEE 802.15.4d-2009, IEEE Standards Association - WPAN Working Group (2009)

94. Part 15.4: Wireless medium access control (MAC) and physical layer (PHY) specifications for low rate wireless personal area networks (WPANs): active radio frequency identification (RFID) system physical layer (PHY). Technical Report IEEE 802.15.4f-2012, IEEE Standards Association - WPAN Working Group (2012)
95. Part 15.4: Wireless medium access control (MAC) and physical layer (PHY) specifications for low rate wireless personal area networks (WPANs): MAC sublayer. Technical Report IEEE 802.15.4e-2012, IEEE Standards Association - WPAN Working Group (2012)
96. Part 15.4: Wireless medium access control (MAC) and physical layer (PHY) specifications for low rate wireless personal area networks (WPANs): physical layer (PHY) specifications for low-data-rate, wireless, smart metering utility networks. Technical Report IEEE 802.15.4g-2012, IEEE Standards Association - WPAN Working Group (2012)
97. Part 15.4: Wireless medium access control (MAC) and physical layer (PHY) specifications for low rate wireless personal area networks (WPANs). Technical Report IEEE 802.15.4-2015, IEEE Standards Association - WPAN Working Group (2015)
98. Part 15.5: Mesh topology capability in wireless personal area networks (WPANs). Technical Report IEEE 802.15.5-2009, IEEE Standards Association - WPAN Working Group (2009)
99. Part 15.6: Wireless body area networks. Technical Report IEEE 802.15.6-2012, IEEE Standards Association - WPAN Working Group (2012)
100. Part 15.7: Short-range wireless optical communication using visible light. Technical Report IEEE 802.15.7-2011, IEEE Standards Association - WPAN Working Group (2011)
101. Pattara-Atikom, W., Krishnamurthy, P., Banerjee, S.: Distributed mechanisms for quality of service in wireless LANs. IEEE Wirel. Commun. 10(3), 26–34 (2003)
102. Pavon, J.D.P., Shankar, N.S., Gaddam, V., Challapali, K., Chou, C.T.: The MBOA-WiMedia specification for ultra wideband distributed networks. IEEE Commun. Mag. 44(6), 128–134 (2006)
103. Personal digital cellular telecommunication system. Technical Report RCR STD-27 (2008)
104. Personal handy phone system. Techical Report RCR STD-28 (2011)
105. Qiao, D., Shin, K.G.: Achieving efficient channel utilization and weighted fairness for data communications in IEEE 802.11 WLAN under the DCF. In: Proceedings of IEEE International Workshop on Quality of Service (IWQoS), Miami, FL, pp. 227–236 (2002)
106. Qiu, X., Li, V.O.K.: Dynamic reservation multiple access (DRMA): a new multiple access scheme for personal communication system (PCS). Wirel. Netw. 2(2), 117–128 (1996)
107. Qiu, R.C., Liu, H., Shen, X.: Ultra-wideband for multiple access communications. IEEE Commun. Mag. 43(2), 80–87 (2005)
108. Ramaiyan, V., Kumar, A., Altman, E.: Fixed point analysis of single cell IEEE 802.11e WLANs: uniqueness and multistability. IEEE/ACM Trans. Netw. 16(5), 1080–1093 (2008)
109. Roberts, J., Mocci, U., Virtamo, J.: Broadband Network Teletraffic. Final Report of Action COST 242. Springer, Berlin (1996)
110. Robinson, J.W., Randhawa, T.S.: Saturation throughput analysis of IEEE 802.11e enhanced distributed coordination function. IEEE J. Sel. Areas Commun. 22(5), 917–928 (2004)
111. Roy, S., Foerster, J.R., Somayazulu, V.S., Leeper, D.G.: Ultrawideband radio design: the promise of high-speed, short-range wireless connectivity. Proc. IEEE 92(2), 295–311 (2004)
112. Ruby, R., Pan, J.: Video streaming with PCA and hard vs soft DRP. In: Proceedings of IEEE Global Communications Conference (GLOBECOM), Miami, FL, pp. 1–6 (2010)
113. Ruiz, J.A., Shimamoto, S.: Novel communication services based on human body and environment interaction: applications inside trains and applications for handicapped people. In: Proceedings of IEEE Wireless Communications and Networking Conference (WCNC), Las Vegas, NV, pp. 2240–2245 (2006)
114. Saberinia, E., Tewfik, A.H.: Pulsed and non-pulsed OFDM ultra wideband wireless personal area networks. In: Proceedings of IEEE Ultra Wideband Systems and Technologies (UWBST), Reston, VA, pp. 275–279 (2002)
115. Sadri, A.: Summary of usage models for 802.15. 3c. Technical Report 15, IEEE P802 (2006)
116. Saleh, A.A., Valenzuela, R.A.: A statistical model for indoor multipath propagation. IEEE J. Sel. Areas Commun. 5(2), 128–137 (1987)

117. Shah, R.C., Yarvis, M.D.: Characteristics of on-body 802.15.4 networks. In: Proceedings of IEEE Workshop on Wireless Mesh Networks (WiMesh), Reston, VA, pp. 138–139 (2006)
118. Shen, X., Zhuang, W., Jiang, H., Cai, J.: Medium access control in ultra-wideband wireless networks. IEEE Trans. Veh. Technol. **54**(5), 1663–1677 (2005)
119. Sheu, S.T., Sheu, T.F.: A bandwidth allocation/sharing/extension protocol for multimedia over IEEE 802.11 ad hoc wireless LANs. IEEE J. Sel. Areas Commun. **19**(10), 2065–2080 (2001)
120. Smith, D., Miniutti, D., Hanlen, L.: Characterization of the body-area propagation channel for monitoring a subject sleeping. IEEE Trans. Antennas Propag. **59**(11), 4388–4392 (2011)
121. Stavrakakis, I., Kazakos, D.: A multiuser random-access communication system for users with different priorities. IEEE Trans. Commun. **39**(11), 1538–1541 (1991)
122. Takizawa, K., Aoyagi, T., Kohno, R.: Channel modeling and performance evaluation of UWB-based wireless body area networks. In: Proceedings of IEEE International Conference on Communications (ICC), Dresden, pp. 1–5 (2009)
123. Tasaka, S., Hayashi, K., Ishihashi, Y.: Integrated video and data transmission in the TDD ALOHA-reservation wireless LAN. In: Proceedings of IEEE International Conference on Communications (ICC), Seattle, WA, vol. 3, pp. 1387–1393 (1995)
124. Telecommunications and information exchange between systems – Local and metropolitan area networks – Specific requirements – Part 11: Wireless LAN Medium Access Control (MAC) and Physical Layer (PHY) Specifications. Technical Report ISO/IEC 8802-11:2005 (IEEE Std 802.11-2003 Edition), IEEE Standards Association – Wireless LAN Working Group (2005)
125. TG3c channel modeling sub-committee final report. Technical Report IEEE802.15-07-0584-00-003c, IEEE P802.15 Working Group for Wireless Personal Area Networks (2007)
126. Tickoo, O., Sikdar, B.: Modeling queueing and channel access delay in unsaturated IEEE 802.11 random access MAC based wireless networks. IEEE/ACM Trans. Netw. **16**(4), 878–891 (2008)
127. Tinnirello, I., Bianchi, G.: Rethinking the IEEE 802.11e EDCA performance modeling methodology. IEEE/ACM Trans. Netw. **18**(2), 540–553 (2010)
128. Triple-play services quality of experience (QoE) requirements. Technical Report TR-126, DSL Forum Architecture & Transport Working Group (2006)
129. Ullah, S., Mohaisen, M., Alnuem, M.A.: A review of IEEE 802.15.6 MAC, PHY, and security specifications. Hindawi Int. J. Distrib. Sens. Netw. **9**(4), 1–12 (2013)
130. Vaidya, N.H., Bahl, P., Gupta, S.: Distributed fair scheduling in a wireless LAN. In: Proceedings of ACM International Conference on Mobile Computing and Networking (MOBICOM), Boston, MA, pp. 167–178 (2000)
131. Veres, A., Campbell, A.T., Barry, M., Sun, L.H.: Supporting service differentiation in wireless packet networks using distributed control. IEEE J. Sel. Areas Commun. **19**(10), 2081–2093 (2001)
132. Wang, P., Zhuang, W.: A collision-free MAC scheme for multimedia wireless mesh backbone. IEEE Trans. Wirel. Commun. **8**(7), 3577–3589 (2009)
133. Wilkinson, R.: Theories for toll traffic engineering in the U.S.A. Bell Syst. Tech. J. **35**(2), 421–514 (1956)
134. Wilson, P., Johnstone, M., Neely, M., Boles, D.: Dynamic storage allocation: a survey and critical review. In: Memory Management, pp. 1–116. Springer, Berlin (1995)
135. Wimedia logical link control protocol. Technical Report, WiMedia Alliance (2007)
136. Win, M.Z., et al.: Ultra-wide bandwidth time-hopping spread-spectrum impulse radio for wireless multiple-access communications. IEEE Trans. Commun. **48**(4), 679–689 (2000)
137. Winands, E., Denteneer, T., Resing, J., Rietman, R.: A finite-source feedback queueing network as a model for the IEEE 802.11 DCF. Eur. Trans. Telecommun. **16**(1), 77–89 (2005)
138. Wong, D., Chin, F., Shajan, M., Chew, Y.: Performance analysis of saturated throughput of PCA in the presence of hard DRPs in wimedia MAC. In: Proceedings of IEEE Wireless Communications and Networking Conference (WCNC), Hong Kong, pp. 423–429 (2007)
139. Wu, H., Peng, Y., Long, K., Cheng, S., Ma, J.: Performance of reliable transports protocol over IEEE 802.11 wireless LAN: analysis and enhancement. In: Proceedings of IEEE/ACM International Conference on Computer Communications (INFOCOM), New York, pp. 599–607 (2002)

140. Wu, H., Xia, Y., Zhang, Q.: Delay analysis of DRP in MBOA UWB MAC. In: Proceedings of IEEE International Conference on Communications (ICC), Istanbul, pp. 229–233 (2006)

141. Xiao, Y.: Backoff-based priority schemes for IEEE 802.11. In: Proceedings of IEEE International Conference on Communications (ICC), Anchorage, AK, pp. 1568–1572 (2003)

142. Xiao, Y.: Enhanced DCF of IEEE 802.11e to support QoS. In: Proceedings of IEEE Wireless Communications and Networking Conference (WCNC), New Orleans, LA, pp. 1291–1296 (2003)

143. Xiao, Y.: Performance analysis of priority schemes for IEEE 802.11 and IEEE 802.11e wireless LANs. IEEE Trans. Wirel. Commun. **4**(4), 1506–1515 (2005)

144. Xiao, Y., Li, H.: Evaluation of distributed admission control for the IEEE 802.11e EDCA. IEEE Commun. Mag. **42**(9), 20–24 (2004)

145. Xu, K., Wang, Q., Hassanein, H.: Performance analysis of differentiated QoS supported by IEEE 802.11e enhanced distributed coordination function (EDCF) in WLAN. In: Proceedings of IEEE Global Communications Conference (GLOBECOM), San Francisco, CA, pp. 1048–1053 (2003)

146. Yang, L., Giannakis, G.B.: Ultra-wideband communications: an idea whose time has come. IEEE Signal Process. Mag. **21**(6), 26–54 (2004)

147. Yu, J.Y., Liao, W.C., Lee, C.Y.: A MT-CDMA based wireless body area network for ubiquitous healthcare monitoring. In: Proceedings of Biomedical Circuits and Systems Conference (BioCAS), London, pp. 98–101 (2006)

148. Zeidler, E.: Nonlinear Functional Analysis and Its Applications, 1: Fixed-Point Theorems. Springer, Berlin (1986)

149. Zhai, H., Kwon, Y., Fang, Y.: Performance analysis of IEEE 802.11 MAC protocols in wirless LANs. Wirel. Commun. Mob. Comput. **4**(8), 917–931 (2004)

150. Zhang, H.: Service disciplines for guaranteed performance service in packet-switching networks. Proc. IEEE **83**(10), 1374–1396 (1995)

151. Zhang, R., Cai, L.: A packet-level model for UWB channel with people shadowing process based on angular spectrum analysis. IEEE Trans. Wirel. Commun. **8**(8), 4048–4055 (2009)

152. Zhang, R., Cai, L.: Joint AMC and packet fragmentation for error-control over fading channels. IEEE Trans. Veh. Technol. **59**(6), 3070–3080 (2010)

153. Zhang, L., Shu, Y., Yang, O., Wang, G.: Study of medium access delay in IEEE 802.11 wireless networks. IEICE Trans. Commun. **89**, 1284–1293 (2006)

154. Zhang, Y., Bin, L., Qi, C.: Characterization of on-human-body UWB radio propagation channel. Microw. Opt. Technol. Lett. **49**(6), 1356–1371 (2007)

155. Zhang, R., Ruby, R., Pan, J., Cai, L., Shen, X.: A hybrid reservation/contention-based MAC for video streaming over wireless networks. IEEE J. Sel. Areas Commun. **28**(3), 389–398 (2010)

156. Zhang, R., Cai, L., Pan, J.: Performance study of hybrid MAC using soft reservation for wireless networks. In: Proceedings of IEEE International Conference on Communications (ICC), Kyoto, pp. 1–5 (2011)

157. Zhang, R., Cai, L., Pan, J., Shen, X.: Resource management for video streaming in ad hoc networks. Elsevier Ad Hoc Netw. **9**(4), 623–634 (2011)

158. Zhao, J., Guo, Z., Zhang, Q., Zhu, W.: Performance study of MAC for service differentiation in IEEE 802.11. In: Proceedings of IEEE Global Communications Conference (GLOBECOM), Taipei, pp. 778–782 (2002)

159. Zheng, Y., Lu, K., Wu, D., Fang, Y.: Performance analysis of IEEE 802.11 DCF in imperfect channels. IEEE Trans. Veh. Technol. **55**(5), 1648–1656 (2006)

160. Zhu, H., Chlamtac, I.: An analytical model for IEEE 802.11e EDCF differential services. In: Proceedings of International Conference on Computer Communications and Networks (ICCCN), Dallas, TX, pp. 163–168 (2003)

161. Zhu, H., Chlamtac, I.: Performance analysis for IEEE 802.11e EDCF service differentiation J. IEEE Trans. Wirel. Commun. **4**(4), 1779–1788 (2005)

162. Zhuang, W., Shen, X., Bi, Q.: Ultra-wideband wireless communications. Wirel. Commun. Mob. Comput. **3**(6), 663–685 (2003)

Printed in the United States
By Bookmasters